Metal Additive Manufacturing: Design, Performance, and Applications

Metal Additive Manufacturing: Design, Performance, and Applications

Guest Editors

Nikolaos Kladovasilakis
Konstantinos Tsongas
Dimitrios Tzetzis

MDPI

Basel • Beijing • Wuhan • Barcelona • Belgrade • Novi Sad • Cluj • Manchester

Guest Editors

Nikolaos Kladovasilakis
Information Technologies Institute
Centre for Research and Technology-Hellas
Thessaloniki
Greece

Konstantinos Tsongas
Department of Industrial Engineering and Management
International Hellenic University
Thessaloniki
Greece

Dimitrios Tzetzis
School of Science and Technology
International Hellenic University
Thessaloniki
Greece

Editorial Office
MDPI AG
Grosspeteranlage 5
4052 Basel, Switzerland

This is a reprint of the Special Issue, published open access by the journal *Materials* (ISSN 1996-1944), freely accessible at: https://www.mdpi.com/journal/materials/special_issues/IH48K0EQR0.

For citation purposes, cite each article independently as indicated on the article page online and as indicated below:

Lastname, A.A.; Lastname, B.B. Article Title. *Journal Name* **Year**, *Volume Number*, Page Range.

ISBN 978-3-7258-3789-2 (Hbk)
ISBN 978-3-7258-3790-8 (PDF)
https://doi.org/10.3390/books978-3-7258-3790-8

Contents

About the Editors

Nikolaos Kladovasilakis

Dr. Nikolaos Kladovasilakis is a Mechanical Engineer and Designer with extensive experience in industrial design, optimization procedures, and advanced manufacturing processes. He holds an M.Eng. in Mechanical Engineering (AUTH, 2017), an M.Sc. in Strategic Product Design with a specialization in industrial design and innovation (IHU, 2020), and a Ph.D. in Material Science and Additive Manufacturing (IHU, 2023). His research focuses on advanced manufacturing technologies, material characterization, process optimization, and computational mechanics. Dr. Kladovasilakis is involved as a senior researcher in numerous projects related to additive manufacturing, robotics, ICT, reverse engineering, circular economy, computational mechanics, and material characterization. He is the author of several scientific publications and holds multiple patents as an inventor.

Konstantinos Tsongas

Dr. Konstantinos Tsongas is a Professor at the International Hellenic University, specializing in advanced materials and their applications in mechanical constructions. He holds an M.Eng. in Civil Engineering (DUTH, 2011), an M.Sc. in Structural Engineering (DUTH, 2012), and a Ph.D. in Mechanical Engineering (AUTH, 2016), with expertise in nanocomposite polymers and the dynamic mechanical behavior of structures. His research interests span additive manufacturing, reverse engineering, mechanical testing, material morphology characterization (SEM/EDS and XRD), FEM, and optimization processes. Dr. Tsongas has actively contributed to numerous research programs related to digital manufacturing and advanced construction and has published extensively in prestigious scientific journals.

Dimitrios Tzetzis

Dr. Dimitrios Tzetzis is a Professor at the International Hellenic University, specializing in mechanical design (CAD/CAE), additive manufacturing, and materials engineering. He is also Head of the Digital Manufacturing and Materials Characterization Laboratory (DMMC Lab) and Director of the M.Sc. in Strategic Product Design at the School of Science and Technology, IHU. His research interests focus on composites and nanocomposites, the processing and mechanics of materials, destructive and non-destructive testing, product design, 3D printing, and nanomaterials. He is actively involved in various funded research and development projects and has published a multitude of scientific studies in the field of advanced materials and structures for various applications.

materials MDPI

Editorial

Metal Additive Manufacturing: Design, Performance, and Applications

Nikolaos Kladovasilakis [1,*], Konstantinos Tsongas [2,*] and Dimitrios Tzetzis [3,*]

1 Centre for Research and Technology Hellas, Information Technologies Institute (CERTH/ITI), 57001 Thessaloniki, Greece
2 Advanced Materials and Manufacturing Technologies Laboratory, Department of Industrial Engineering and Management, School of Engineering, International Hellenic University, 57001 Thessaloniki, Greece
3 Digital Manufacturing and Materials Characterization Laboratory, School of Science and Technology, International Hellenic University, 14th Km Thessaloniki-Moudania, Thermi, 57001 Thessaloniki, Greece
* Correspondence: nikoklad@iti.gr (N.K.); ktsongas@ihu.gr (K.T.); d.tzetzis@ihu.edu.gr (D.T.)

The rapid evolution of metal additive manufacturing (AM) has led to transformative advancements across various industries, including aerospace, healthcare, automotive, and energy. While this groundbreaking technology holds immense promise, its widespread adoption faces significant challenges that demand innovative solutions in design, performance optimization, and application-specific development. The Special Issue titled "Metal Additive Manufacturing: Design, Performance, and Applications" addresses these challenges by serving as a dedicated platform for disseminating cutting-edge research and pioneering insights. By presenting novel methodologies, rigorously evaluating material properties, and showcasing state-of-the-art real-life applications, this collection of high-quality scientific papers aims to deepen our understanding and accelerate the implementation of metal AM in industry.

This Special Issue comprises 10 high-quality papers authored by 47 contributors from 22 distinguished institutes and organizations worldwide, reflecting the collaborative and interdisciplinary nature of this Special Issue. The studies featured in this Special Issue build on a foundation of 381 conducted and published works, culminating in comprehensive and novel findings that advance the field. The significant response from the scientific community further underscores the significance of this Special Issue, with over 9600 reads recorded within the first nine months of its release.

Received: 31 December 2024
Accepted: 22 January 2025
Published: 25 January 2025

Citation: Kladovasilakis, N.; Tsongas, K.; Tzetzis, D. Metal Additive Manufacturing: Design, Performance, and Applications. *Materials* **2025**, *18*, 545. https://doi.org/10.3390/ma18030545

The published papers in this Special Issue were balanced between metal AM product design, performance, and applications. In detail, Liu H. et al. [1] contributed a paper entitled "Thermal behavior and mechanical properties of different lattice structures fabricated by selective laser melting", providing research findings on the design and performance of metal AM. Xia P. et al. [2] contributed a paper entitled "Study on Microstructure and Thermal Cracking Sensitivity of Deposited Ti6Al4V/Inconel 718 Composites Made by Two-Wire Arc Additive Manufacturing by Current", providing research findings on the performance of metal AM. Lee J. et al. [3] contributed a paper entitled "Study On Surface Quality Analysis Of An Uncoated Boron Steel And Its Oxide Layer Suppression Method For Hot Stamping", providing research findings on the performance of metal AM. Perrone A. et al. [4] contributed a paper entitled "Various Configurations for Improving the Efficiency of Metallic and Superconducting Photocathodes Prepared by Pulsed Laser Deposition: A Comparative Review", providing research findings on the performance and applications of metal AM. Giarmas E. et al. [5] contributed a paper entitled "Nanoindentation Creep Behavior of Additively Manufactured H13 Steel by Utilizing Selective Laser Melting

Technology", providing research findings on performance and applications of metal AM. Egorov S. et al. [6] contributed a paper entitled "Effect of the Cooling Liquid on the Milled Interface in the Combined Process of Milling and Direct Metal Deposition", providing research findings on performance and applications of metal AM. Kim S. et al. [7] contributed a paper entitled "Accuracy of Mandibular Removable Partial Denture Frameworks Fabricated by 3D Printing and Conventional Techniques", providing research findings on the performance and applications of metal AM. Kladovasilakis N. et al. [8] contributed a paper entitled "Metal 3D-Printed Bioinspired Lattice Elevator Braking Pads for Enhanced Dynamic Friction Performance", providing research findings in the design and applications of metal AM. Zykona A. et al. [9] contributed a paper entitled "Effect of "ColdArc" WAAM Regime and Arc Torch Weaving on Microstructure and Properties of As-Built and Subtransus Quenched Ti-6Al-4V", providing research findings in applications of metal AM. Finally, Hawryluk M. et al. [10] contributed a paper entitled "Improvement of the technology of precision forging of connecting rod-type forgings in multiple systems, in the aspect of the possibilities of process robotization by means of numerical modeling", providing research findings on the performance and applications of metal AM.

This Special Issue not only advances the scientific understanding of metal AM but also serves as a bridge between research innovation and industrial application. We extend our sincere gratitude to all the authors who contributed their exceptional work to this Special Issue. Their valuable research and dedication have been instrumental in shaping the success of this collection. We also thank the reviewers for their invaluable expertise in ensuring the quality and rigor of the published studies. Finally, we thank the editorial team for their unwavering support and commitment throughout this process. Together, these efforts have created a meaningful contribution to the field and will undoubtedly inspire further advancements in metal AM.

Author Contributions: N.K., K.T. and D.T. equally contributed to this research. All authors have read and agreed to the published version of the manuscript.

Conflicts of Interest: The authors declare no conflicts of interest.

References

1. Liu, H.; Cai, G.; Peng, K.; Jin, H.; Alexander, A. Thermal Behavior and Mechanical Properties of Different Lattice Structures Fabricated Using Selective Laser Melting. *Materials* **2024**, *17*, 5603. [CrossRef]
2. Xia, P.; Ye, X.; Zhang, G. Study on Microstructure and Thermal Cracking Sensitivity of Deposited Ti6Al4V/Inconel 718 Composites Made by Two-Wire Arc Additive Manufacturing by Current. *Materials* **2024**, *17*, 5989. [CrossRef]
3. Lee, J.; Song, J.; Bae, G. Study on Surface Quality Analysis of an Uncoated Boron Steel and Its Oxide Layer Suppression Method for Hot Stamping. *Materials* **2024**, *17*, 5563. [CrossRef] [PubMed]
4. Perrone, A.; Aziz, M.R.; Gontad, F. Various Configurations for Improving the Efficiency of Metallic and Superconducting Photocathodes Prepared by Pulsed Laser Deposition: A Comparative Review. *Materials* **2024**, *17*, 5257. [CrossRef] [PubMed]
5. Giarmas, E.; Tzimtzimis, E.K.; Kladovasilakis, N.; Tzovaras, D.; Tzetzis, D. Nanoindentation Creep Behavior of Additively Manufactured H13 Steel by Utilizing Selective Laser Melting Technology. *Materials* **2024**, *17*, 3756. [CrossRef] [PubMed]
6. Egorov, S.; Schudeleit, T.; Wegener, K. Effect of the Cooling Liquid on the Milled Interface in the Combined Process of Milling and Direct Metal Deposition. *Materials* **2024**, *17*, 3119. [CrossRef] [PubMed]
7. Kim, S.; Oh, K.C.; Kim, J.-H. Accuracy of Mandibular Removable Partial Denture Frameworks Fabricated by 3D Printing and Conventional Techniques. *Materials* **2024**, *17*, 3148. [CrossRef] [PubMed]
8. Kladovasilakis, N.; Pechlivani, E.M.; Sfampa, I.K.; Tsongas, K.; Korlos, A.; David, C.; Tzovaras, D. Metal 3D-Printed Bioinspired Lattice Elevator Braking Pads for Enhanced Dynamic Friction Performance. *Materials* **2024**, *17*, 2765. [CrossRef] [PubMed]
9. Zykova, A.; Savchenko, N.; Nikolaeva, A.; Panfilov, A.; Vorontsov, A.; Semenchuk, V.; Gurianov, D.; Kolubaev, E.; Tarasov, S. Effect of "ColdArc" WAAM Regime and Arc Torch Weaving on Microstructure and Properties of As-Built and Subtransus Quenched Ti-6Al-4V. *Materials* **2024**, *17*, 2325. [CrossRef] [PubMed]

10. Hawryluk, M.; Dudkiewicz, Ł.; Polak, S.; Barełkowski, A.; Miżejewski, A.; Szymańska, T. Improvement of the Technology of Precision Forging of Connecting Rod-Type Forgings in a Multiple System, in the Aspect of the Possibilities of Process Robotization by Means of Numerical Modeling. *Materials* **2024**, *17*, 1087. [CrossRef] [PubMed]

materials

Article

Thermal Behavior and Mechanical Properties of Different Lattice Structures Fabricated Using Selective Laser Melting

Hui Liu [1], Gaoshen Cai [1,2,*], Kai Peng [1], Haozhe Jin [1,3,*] and Antonov Alexander [4]

[1] School of Mechanical Engineering, Zhejiang Sci-Tech University, Hangzhou 310018, China
[2] Key Laboratory of Advanced Manufacturing Technology of Zhejiang Province, School of Mechanical Engineering, Zhejiang University, Hangzhou 310027, China
[3] Institute of Flow-Induced Corrosion, Zhejiang Sci-Tech University, Hangzhou 310018, China
[4] Innovative Technologies of Mechanical Engineering Faculty, Yanka Kupala State University of Grodno, 230023 Grodno, Belarus
* Correspondence: caigaocan@zstu.edu.cn (G.C.); haozhejin@zstu.edu.cn (H.J.)

Abstract: In this study, the size of molten pool and the porosity of parts under different processing parameters are studied using numerical simulation. According to the results, the appropriate processing parameters were selected to simulate the temperature and residual stress distribution during the forming process of body-centered cube (BCC), face-centered cube (FCC) and rhombic dodecahedron (Dode) lattice structures. In addition, three lattice structures were fabricated via selective laser melting (SLM) technology, and quasi-static compression experiments were carried out to study their mechanical properties. The results show that the high temperature parts of the three structures are all under the node and their adjacent pillars, and the closer to the nodes, the higher the temperature. The residual stress of the Dode structure is the highest, reaching 1218.2 MPa. It is also found that the residual stress in the Z direction is the largest, which plays a dominant role in the forming process. Through compression experiments, it is found that diagonal shear failure occurs in all three lattice structures, and Dode shows the best compression performance.

Keywords: selective laser melting; residual stress; numerical simulation; lattice structure; mechanical property

Citation: Liu, H.; Cai, G.; Peng, K.; Jin, H.; Alexander, A. Thermal Behavior and Mechanical Properties of Different Lattice Structures Fabricated Using Selective Laser Melting. *Materials* **2024**, *17*, 5603. https://doi.org/10.3390/ma17225603

Academic Editors: Dimitrios Tzetzis, Nikolaos Kladovasilakis and Konstantinos Tsongas

Received: 28 October 2024
Revised: 8 November 2024
Accepted: 13 November 2024
Published: 16 November 2024

1. Introduction

Titanium alloys have the advantages of low density, corrosion resistance, excellent fatigue strength and heat resistance [1]. Due to these characteristics, titanium alloys have great potential as lightweight structural materials. The excellent properties of Ti6Al4V make it widely used, such as gas turbines, automotive parts and medical implants and prostheses, which are welcomed by many industries [2,3].

A lattice structure has the advantages of light weight, high specific strength and great heat dissipation. It has attracted much attention in aerospace, biomedical and other fields [4–6]. The traditional manufacturing methods of a lattice structure include investment casting [7], snap fit, expanded metal sheet, metallic wire assembly [8], etc. However, it is difficult to manufacture geometrically complex parts using traditional manufacturing techniques. Additive manufacturing technology can overcome this difficulty and realize the manufacture of complex structures.

Selective laser melting (SLM) is a common metal additive manufacturing technology [9,10]. SLM technology uses a laser as a heat source to selectively melt the powder according to the shape of the model, and the powder is repeatedly melted until the part is manufactured [11,12]. Although the preparation of lattice structures using SLM technology has a good prospect, defects will occur due to the limitations of the process itself [13]. In addition, the powder particles undergo two processes of rapid melting and rapid solidification in the SLM forming process, resulting in large thermal stress and temperature difference, resulting in large residual stress and warpage deformation, affecting the accuracy and mechanical properties [14,15]. In order to improve the forming quality, it is necessary to study the residual stress.

Zhao et al. [16] studied the influence of different processing parameters on the residual stress of SLM-formed parts. It is found that the size and temperature gradient of molten pool are affected by the change of processing parameters. The larger the temperature gradient of the molten pool, the larger the residual stress, and the maximum residual stress there is in the overlapping area of the molten pool. The research of Wang et al. [17] showed that laser power, scanning speed and scanning strategy have different effects on the residual stress of parts, and the distribution of residual stress can be controlled by changing the processing parameters. In addition, it has also been found that there is a heat accumulation effect in the process of SLM manufacturing parts, and the existence of the heat accumulation effect makes the residual stress increase. Gan et al. [18] found that the angle and diameter of the lattice structure pillar significantly affect the residual stress. The residual stress can be reduced by increasing the inclination angle and decreasing the diameter of the pillar. The change of pillar length has little effect on the residual stress.

The measurement methods of residual stress include the drilling method, neutron diffraction method, X-ray method, etc. [19]. The drilling method requires holes to be drilled on the surface of the component, which is difficult to achieve for a lattice structure. Tobias et al. [20] introduced a non-destructive neutron diffraction method to measure the residual stress in lattice structures. However, they proposed that due to the particularity of the lattice structure, it is difficult to obtain accurate results and prone to errors.

At present, there are relatively few studies on the thermal behavior of lattice structures. Therefore, several types of lattice structures such as BCC, FCC and Dode were selected in this study, and the temperature change and residual stress distribution of the Ti6Al4V lattice structure were systematically studied. The parts were prepared using SLM technology, and their compression response and failure behavior were studied.

2. Materials and Methods

2.1. Materials and Manufacturing

The raw material is Ti6Al4V powder produced by Bright Laser Technologies Co., Ltd. (Xi'an, China). The chemical composition of the powder is listed in Table 1. JSM-5610LV scanning electron microscopy (SEM) images show that the morphology of the observed powder particles is close to spherical, as shown in Figure 1.

Table 1. Chemical compositions (wt.%).

Alloying Elements	Ti	Al	V	Fe	C	N	H	O
wt.%	Bal.	5.5–6.75	3.5–4.5	≤0.30	≤0.08	≤0.05	≤0.015	0.08–0.13

Bal.: Balance, the amount of remaining ingredients.

The lattice structures were fabricated using the company's BLT-A300 + 3D printer using laser power of 300 W, scanning speed of 1000 mm/s, scanning spacing of 0.1 mm and layer thickness of 0.03 mm, as shown in Figure 2.

In order to test the mechanical properties of the lattice structures, a quasi-static compression experiment was carried out using a microcomputer-controlled electronic universal testing machine (manufactured by Ji'nan Tenson Testing Machine Manufacturing Co., Ltd., Jinan, China) with a maximum load of 100 kN, and the compression was carried out at a speed of 1 mm/min until the part failed. Then, the fracture morphology was observed via SEM, and the fracture mechanism of the samples were analyzed.

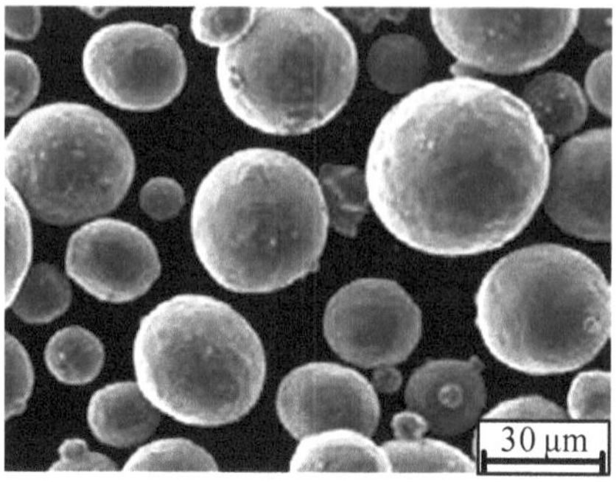

Figure 1. SEM image of Ti6Al4V powder.

Figure 2. Lattice structure fabricated using SLM.

2.2. Numerical Simulation

The thermal behavior of SLM is very complex. To obtain accurate simulation results, it is necessary to have the temperature control equation, heat source model, initial conditions and boundary conditions. In this study, an Nd: YAG fiber laser was used for SLM, and the laser energy density was approximately Gaussian distribution. The Gaussian heat source function is shown in Equation (1) [21]:

$$q = \frac{2AP}{\pi R_L^2} \exp(-\frac{2r^2}{R_L^2}) \tag{1}$$

where A is the laser absorption rate of the powder, P is the laser power, R_L is the radius of the laser spot, and r is the radial distance from the node to the center of the laser spot.

In addition, the correct initial conditions and boundary conditions should be set. Equation (2) is the initial condition.

$$T|_{t=0} = T_0(x, y, z, t) \tag{2}$$

where T is the ambient temperature.

The boundary conditions in the numerical simulation model can be expressed using Equation (3) and (4), respectively [22]:

$$-\lambda_e \frac{\partial \mathrm{T}}{\partial \mathrm{n}} = \mathrm{h}(\mathrm{T} - \mathrm{T}_m) \tag{3}$$

$$-\lambda_e \frac{\partial T}{\partial n} = \sigma_{SB}\psi(T^2 + T_m^2)(T + T_m)(T - T_m) = \beta_0(T - T_m) \tag{4}$$

where λ_e is the effective thermal conductivity of the powder, n is the vector of the top of the powder bed along the normal direction, h is the convective heat transfer coefficient, T_m

is the ambient temperature, σ_{SB} is the Stefan-Boltzman constant, ψ is the reflectivity of the material, and β_0 is the thermal radiation coefficient.

The influence of temperature change on the thermal physical property parameters of Ti6Al4V material used in the numerical simulation is shown in Table 2 [23].

Table 2. The mechanical property of Ti6Al4V.

Temperature (K)	Coefficient of Thermal Expansion ($\times 10^6$/K)	Modulus of Elasticity (GPa)	The Yield Strength (MPa)	Poisson's Ratio	Shear Modulus (MPa)
298	8.91	107	1098	0.323	0.7
589	10.12	91.8	844	0.339	2.2
700	10.82	82.4	663	0.347	2.2
811	11.31	69.6	527	0.354	1.9
923	11.71	54.7	300	0.361	1.9
1073	12.21	35.3	60	0.369	2
1123	12.37	29.3	5	0.371	2
1473	12.44	6.6	2	0.392	2
1923	12.51	0.1	0.1	0.415	0.1

NX was used to model the lattice structure, as shown in Figure 3; the size of each unit was $5 \times 5 \times 5$, and the diameter of the pillar is 1 mm. The model was imported into ANSYS Workbench for simulation. The material is Ti6Al4V powder. The processing parameters are the same as the actual manufacturing. Tetrahedral mesh with a unit length of 0.5 mm was used.

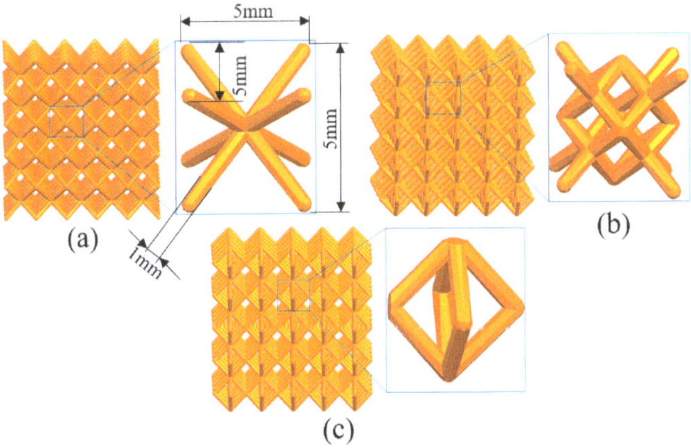

Figure 3. Lattice structures: (**a**) BCC, (**b**) Dode and (**c**) FCC.

3. Optimization of SLM Processing Parameters

3.1. The Influence of Processing Parameters on the Molten Pool

The schematic diagram of the molten pool shape is shown in Figure 4. The shape of the molten pool is greatly affected by the process parameters, which are mainly reflected in the width and depth of the molten pool. Therefore, the same method as the previous study was used to study the influence of process parameters on the molten pool geometry [24,25].

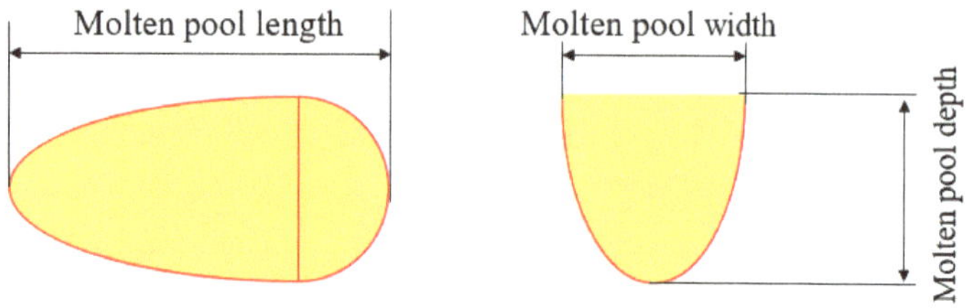

Figure 4. Molten pool shape.

Through the combination of different processing parameters, a variety of combinations are obtained. The influence of processing parameters on the molten pool is shown in Figure 5. With the increase of laser power, the depth, width and length of the molten pool increase. With the increase of scanning speed, the depth and width of the molten pool decrease, and the length of the molten pool decreases first and then increases. In the study of Chen [26] and Zhao et al. [27], the influence of process parameters on the size of the molten pool is consistent with the rule in this study.

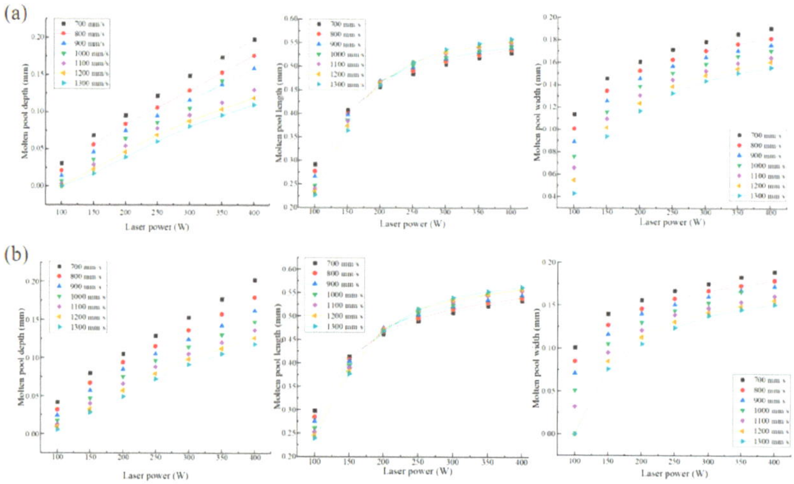

Figure 5. Molten pool size of different processing parameters: (**a**) 0.03 mm layer thickness and (**b**) 0.04 mm layer thickness.

The remelting densification mechanism is proposed as a method to eliminate residual pores. When the current powder layer is melted, sufficient energy input allows the previous layer of powder to melt again, which is beneficial to reduce the porosity of the previous layer [28]. Therefore, in order to obtain a good molten pool, at least a depth of 2.5 layers of molten pool is required, because the penetration depth of 2.5 layers can reduce the porosity by remelting the previous layer. The molten pool morphology of different process parameters is shown in Figure 6. Different patterns represent different weld pool sizes. The orange square indicates that the depth of the molten pool is too shallow, the blue dots indicate that the molten pool is too deep, the yellow triangle indicates that the molten pool is too long, and the green stars indicate that the size of the molten pool is reasonable. Compared with 0.04 mm layer thickness, 0.03 mm layer thickness can produce a more suitable molten pool size.

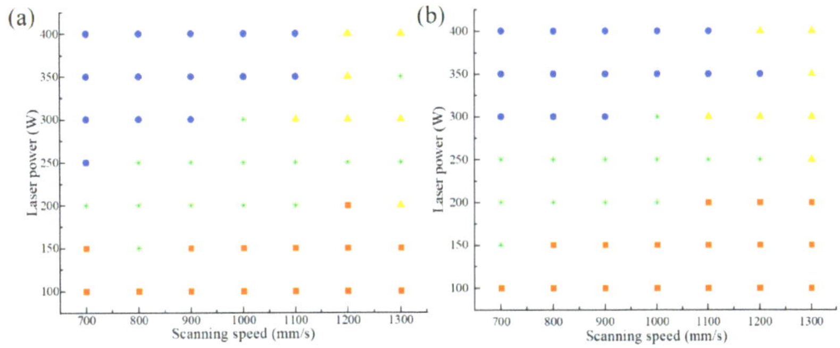

Figure 6. Molten pool quality of different processing parameters: (**a**) 0.03 mm layer thickness and (**b**) 0.04 mm layer thickness.

3.2. The Influence of Processing Parameters on Porosity

According to the simulation results of the molten pool, the combination of the processing parameters in the green part of Figure 6 was selected for porosity simulation. In addition, the new variable scan spacing was added. There are many reasons that affect the porosity. In the manufacturing process, the entrained gas in the powder may remain in the part. Contaminant degassing may occur during the formation of the molten pool. During the solidification process, a small amount of gas vapor may also be retained in the molten pool, forming pores inside the part. The results are shown in Figure 7; when the scanning spacing does not exceed 0.1 mm and the laser power is 300 W, the porosity is low.

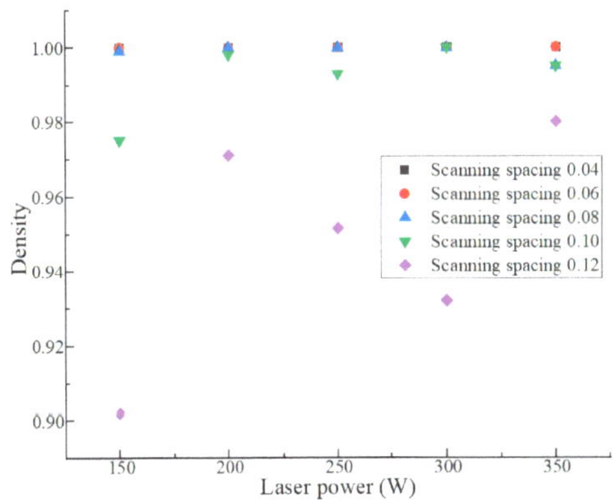

Figure 7. The simulation results of porosity.

Considering the simulation results of molten pool and porosity, the optimized process parameters are as follows: laser power 300 W, scanning speed 1000 mm/s, layer thickness 30 μm, scanning spacing 0.1 mm.

4. Thermal Behavior

4.1. Temperature Distribution of Different Cell Lattice Structures

The temperature change and residual stress of the three lattice structures were simulated using the optimized process parameters. Figure 8 shows the temperature distribution of three different cell structures during SLM forming. The high temperature parts of the three structures are all under the nodes and their adjacent pillars. The closer to the node, the higher the temperature. This is due to the repeated remelting near the node, and the continuous accumulation of temperature making the temperature rise. The maximum temperatures of BCC, Dode and FCC are 1270.77 °C, 1298.77 °C and 1232.39 °C, respectively. Among the three structures, the Dode structure has the highest temperature, mainly because the Dode structure is complex, the number of pillars is more than that of other structures, and the total input of energy in the manufacturing process is high and the heat dissipation is poor, making its highest temperature higher than the other two structures.

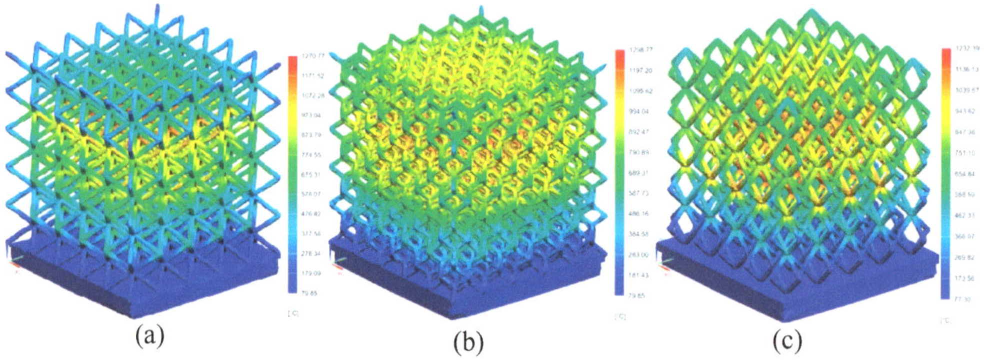

Figure 8. Temperature distribution of different cell lattice structures: (**a**) BCC, (**b**) Dode and (**c**) FCC.

4.2. Residual Stress of Different Cell Lattice Structures

Figure 9 is the residual stress distribution map of different cell lattice structures. It can be found that among the three structures, the equivalent residual stress of the Dode structure is the largest, reaching 1218.2 MPa. The stress of the FCC structure reached 1164.4 MPa, and the stress of the BCC structure reached 1159.9 MPa. The complexity of the Dode structure makes the internal heat dissipation poor, so that it has the greatest residual stress. The maximum residual stress of each structure is at the node, which is because the temperature at the node is higher than at other parts in the forming process (Figure 8), and there is a large temperature gradient.

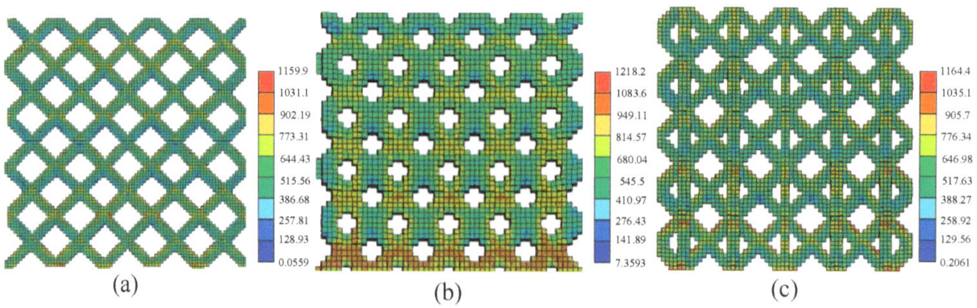

Figure 9. Residual stress distribution of different cell lattice structures: (**a**) BCC, (**b**) Dode and (**c**) FCC.

The stress in different directions was extracted and the stress distribution of the lattice structure was further analyzed. The distribution of residual stress in BCC, Dode and FCC structures is shown in Figure 10. The equivalent residual stress component in Z direction is the largest, and the stress magnitude in other directions is similar. This is due to the excessive temperature gradient at the connection between the substrate and the component. The maximum tensile stresses of BCC, Dode and FCC in the Z direction are 1559.9 MPa, 1818.8 MPa and 2361 MPa, respectively. The maximum compressive stresses are 1330.8 MPa, 1081.0 MPa and 911.16 MPa, respectively.

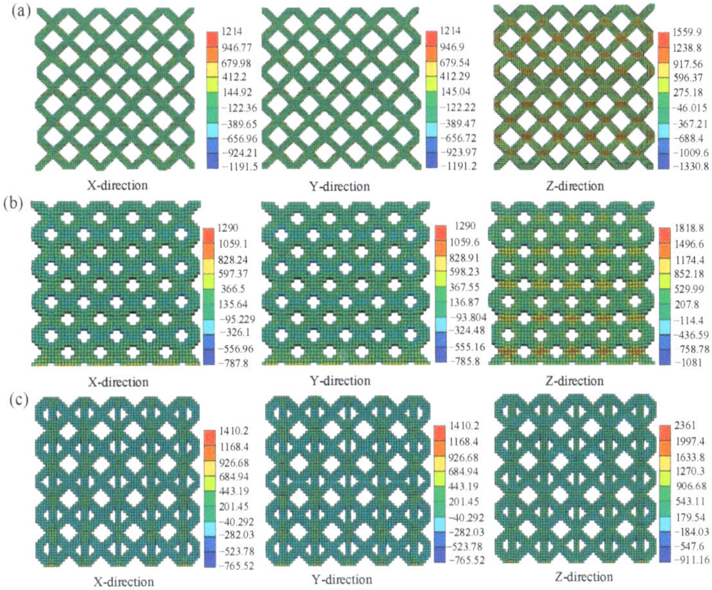

Figure 10. Residual stress distribution in different directions: (**a**) BCC, (**b**) Dode and (**c**) FCC.

5. Results and Discussion

5.1. Quasi-Static Compression Experiment

Figure 11 shows the stress–strain curves of different lattice structures. The compression performance of different lattice structures is very different. Multiple peaks are observed in the stress–strain curve, and the final feature is that the stress drops sharply and the parts are damaged. The slope of the curve represents the elastic modulus, and the first peak stress represents the compressive strength. The slope of the Dode structure in the elastic stage is significantly larger than that of BCC and FCC. The strain at the first peak stress of all the structures is very close, but the peak stress is very different. The mechanical properties of the three lattice structures are shown in Table 3. The Dode structure has the highest elastic modulus and compressive strength, showing the best compression performance. Although FCC has a higher elastic modulus and compressive strength than BCC, it can be seen from Figure 10 that the bearing capacity decreases sharply after the first damage, and begins to be lower than the BCC structure after the strain reaches about 0.14.

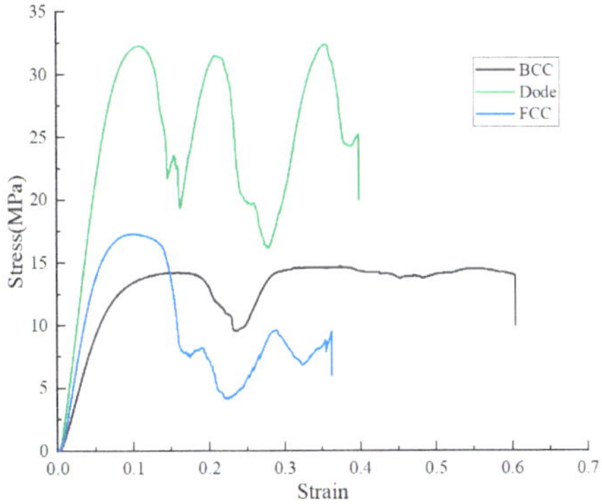

Figure 11. Stress–strain curve of different lattice structures.

Table 3. Mechanical properties of different lattice structures.

Sample	Elastic Modulus (MPa)	Compressive Strength (MPa)	Plateau Stress (MPa)	Yield Strength (MPa)
BCC	208.46	14.2	13.61	10.5
Dode	478.72	32.26	25.69	23.4
FCC	340.96	17.29	8.871	14.59

5.2. Deformation Behavior

Figure 12 shows the deformation of each lattice structure during the compression experiment. In the initial stage of the compression experiment, slight deformation occurred in each structure. As the pressure continues to be applied, the pillars of the bottom unit deform and extend outward, and the bottom deformation of the FCC structure is the most significant. They have different deformation mechanisms after the yield stage. The failure of the three structures occurs at the nodes. The pillars at the fracture of the BCC structure are compressed in a horizontal direction, and the direction of the pillars at the fracture of the FCC structure is almost unchanged. Both are caused by the bending collapse of the pillars. The relative slip of the Dode structure occurs along the 45° direction. Finally, all three structures form extrusion bands along the 45° direction, resulting in catastrophic damage. In the compression process, it is found that the compression performance of the lattice structure substrate will be affected after laser cutting. Laser cutting causes the hardness of the bottom pillar to increase, the plasticity to decrease, and the bottom pillar is more likely to break under load.

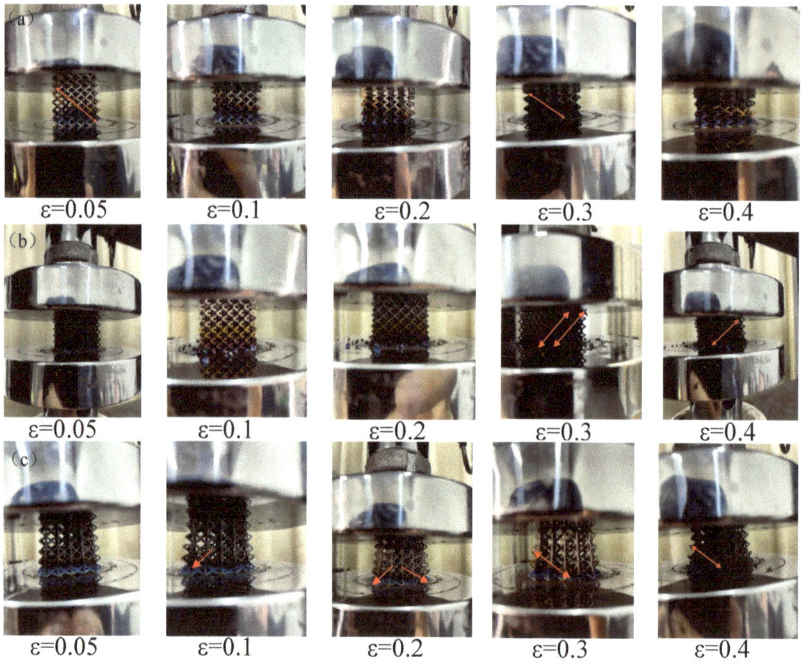

Figure 12. Deformation mode of lattice structure: (**a**) BCC, (**b**) Dode and (**c**) FCC.

5.3. Fracture Morphology

Figure 13a,c,e are the compressive fracture morphology of BCC, Dode and FCC lattice structures, respectively. Figure 13b,d,f correspond to the local amplification diagram, respectively. A large number of dimples are observed in each lattice structure, which is characteristic of the ductile fracture of metals. A ductile fracture is beneficial for the structure to absorb more energy. Secondary cracks and holes are found in Figure 13c. The reason for the secondary crack is that the lattice structure tends to expand outward during the loading process of the compression experiment, resulting in high tension in the weak area in the middle of the pillar, resulting in a mixed fracture. The hole is generated by the gas residue during the manufacturing process. There are more dimples on the BCC fracture than on the FCC fracture, and the dimples on the FCC fracture surface are larger and the tearing is more obvious, indicating that the FCC nodes undergo large plastic deformation before the final fracture.

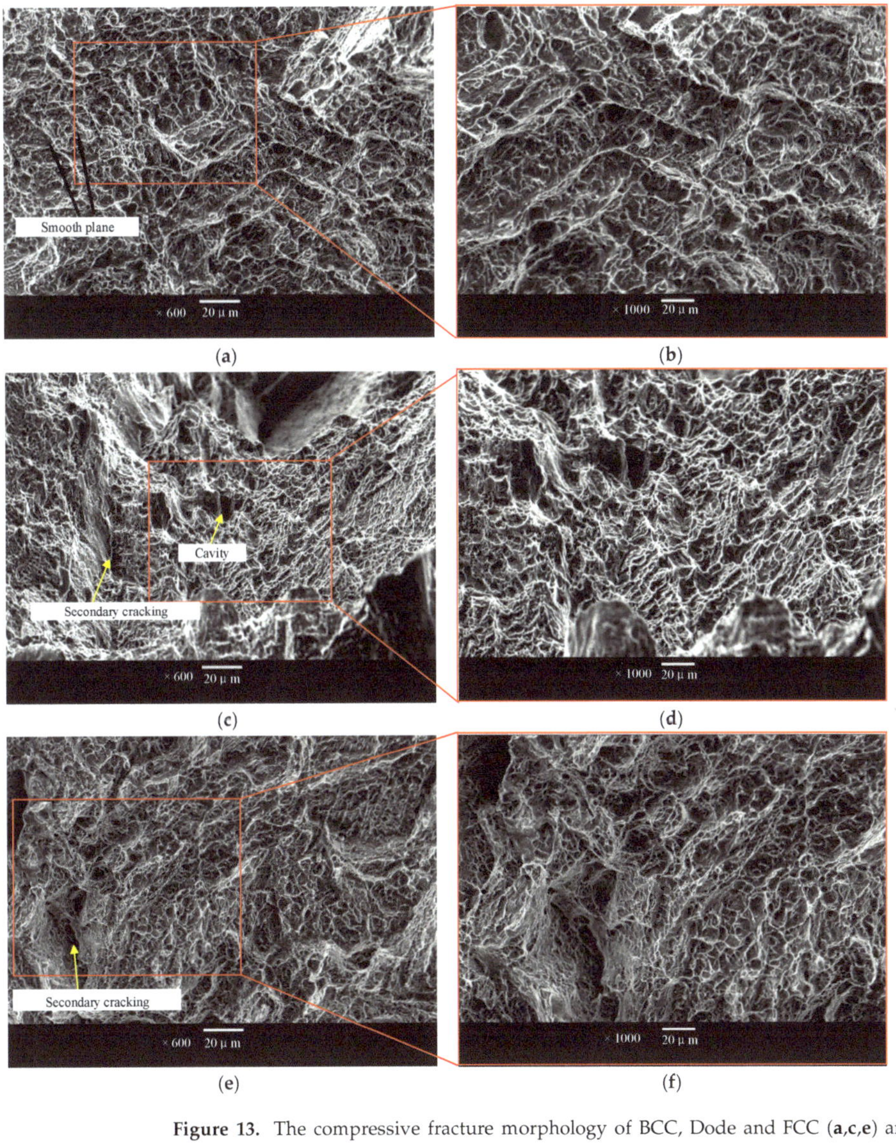

(a) (b) (c) (d) (e) (f)

Figure 13. The compressive fracture morphology of BCC, Dode and FCC (**a**,**c**,**e**) and the local enlargement of the fracture morphology of the lattice structure (**b**,**d**,**f**).

6. Conclusions

In this study, the processing parameters were optimized via numerical simulation. The temperature and residual stress distribution of different cell lattice structures were studied using the optimization results. The compression performance of the lattice structure was tested. The following main conclusions are drawn:

(1) The influence of different processing parameters on the molten pool and porosity were studied, and the optimization was carried out based on this. The optimized parameters were: laser power 300 W, scanning speed 1000 mm/s, powder thickness 0.03 mm and scanning spacing 0.1 mm.
(2) The thermal behavior evolution of the Ti6Al4V lattice structure in the SLM forming process was studied. The high temperature parts of the three structures are all on the lower side of the node and its adjacent pillars, and the closer to the node, the higher the temperature. The temperature of the Dode structure reaches the highest, followed by the BCC structure, and the smallest is the FCC.
(3) The node of the lattice structure is the main area of residual stress distribution, which is closely related to the high temperature distribution at the node. From the residual stress in different directions, the residual compressive stress exists at both ends of the part. The residual stress in the X and Y directions is close, and the residual stress in the Z direction has the greatest influence on the forming process. Moreover, the residual stress of the lattice structure of different cells is significantly different. The residual stress of Dode is the highest, reaching 1218.2 MPa. FCC is the second, and BCC is the lowest.
(4) The compression performance of different cell lattice structures is very different. The Dode structure has the highest elastic modulus and compressive strength, reaching 478.72 MPa and 32.25 MPa, showing the best compression performance among the three.

Author Contributions: Conceptualization, G.C. and H.J.; data curation, H.L. and K.P.; formal analysis, K.P. and G.C.; investigation, H.L.; methodology, K.P and A.A.; project administration, G.C. and H.J.; resources, G.C. and A.A.; software, H.L. and K.P.; validation H.L. and K.P.; visualization K.P.; writing—original draft, H.L. and K.P.; writing—review and editing, G.C. All authors have read and agreed to the published version of the manuscript.

Funding: This work was supported by the Excellent Postgraduate Dissertation Cultivation Foundation of Zhejiang Sci-Tech University, Grant No. LW-YP2024045, Fundamental Research Funds of Zhejiang Sci-Tech University, Grant No. 22242297-Y, Zhejiang Provincial Natural Science Foundation of China, Grant No. LQ18E050010, and China Postdoctoral Science Foundation Funded Project, Grant No. 2018M642482.

Institutional Review Board Statement: Not applicable.

Informed Consent Statement: Not applicable.

Data Availability Statement: The data presented in this study are available on request from the corresponding author.

Conflicts of Interest: The authors declare no conflicts of interest.

References

1. Yang, X.; Ma, W.J.; Zhang, Z.Y.; Liu, S.F.; Tang, H.P. Ultra-high specific strength Ti6Al4V alloy lattice material manufactured via selective laser melting. *Mater. Sci. Eng. A-Struct.* **2022**, *840*, 142956. [CrossRef]
2. Yuan, B.G.; Liu, X.; Du, J.F.; Chen, Q.; Wan, Y.Y.; Xiang, Y.L.; Tang, Y.; Zhang, X.X.; Huang, Z.Y. Effects of hydrogenation temperature on room-temperature compressive properties of CMHT-treated Ti6Al4V alloy. *J. Mater. Sci. Technol.* **2021**, *72*, 132–143. [CrossRef]
3. Bhardwaj, A.; Mudaliar, V.; Gohil, N.; Gupta, A.K.; Kumar, S.S.S. Evolution of microstructure and mechanical properties of Ti6Al4V alloy by multiple passes of constrained groove pressing at elevated temperature. *J. Mater. Process. Tech.* **2021**, *288*, 116891. [CrossRef]

4. Ding, J.H.; Qu, S.; Zhang, L.; Wang, M.Y.; Song, X. Geometric deviation and compensation for thin-walled shell lattice structures fabricated by high precision laser powder bed fusion. *Addit. Manuf.* **2022**, *58*, 103061. [CrossRef]
5. Fan, X.J.; Tang, Q.; Feng, Q.X.; Ma, S.; Song, J.; Jin, M.X.; Guo, F.Y.; Jin, P. Design, mechanical properties and energy absorption capability of graded-thickness triply periodic minimal surface structures fabricated by selective laser melting. *Int. J. Mech. Sci.* **2021**, *204*, 106586. [CrossRef]
6. Pan, C.; Han, Y.F.; Lu, J.P. Design and optimization of lattice structures: A review. *Appl. Sci-Basel.* **2020**, *10*, 6374. [CrossRef]
7. Zhang, L.; Chen, Y.F.; He, R.J.; Bai, X.J.; Zhang, K.Q.; Ai, S.G.; Yang, Y.Z.; Fang, D.N. Bending behavior of lightweight C/SiC pyramidal lattice core sandwich panels. *Int. J. Mech. Sci.* **2020**, *171*, 105409. [CrossRef]
8. Meyer, G.; Wang, H.; Mittelstedt, C. Influence of geometrical notches and form optimization on the mechanical properties of additively manufactured lattice structures. *Mater. Des.* **2022**, *222*, 111082. [CrossRef]
9. Kuai, Z.Z.; Li, Z.H.; Liu, B.; Chen, Y.L.; Lu, S.Y.; Bai, P.K. Effect of heat treatment on CuCrZr alloy fabricated by selective laser melting: Microstructure evolution, mechanical properties and fracture mechanism. *J. Mater. Res. Technol.* **2023**, *23*, 2658–2671. [CrossRef]
10. Jiang, Z.H.; Xu, P.W.; Liang, Y.L.; Liang, Y. Deformation effect of melt pool boundaries on the mechanical property anisotropy in the SLM AlSi10Mg. *Mater. Today. Commun.* **2023**, *36*, 106879. [CrossRef]
11. Sefene, E.M. State-of-the-art of selective laser melting process: A comprehensive review. *J. Manuf. Syst.* **2022**, *63*, 250–274. [CrossRef]
12. Daloadulo, E.; Pini, F.; Leali, F. Powder bed fusion integrated product and process design for additive manufacturing: A systematic approach driven by simulation. *Int. J. Adv. Manuf. Technol.* **2024**, *130*, 5440–5452.
13. Zhang, T.; Zhang, K.F.; Chen, Q. Study on the effects of the processing parameters on the single tracks and the block support structures fabricated by selective laser melting. *J. Laser Appl.* **2024**, *36*, 012024. [CrossRef]
14. Jia, H.L.; Sun, H.; Wang, H.Z.; Wu, Y.; Wang, H.W. Scanning strategy in selective laser melting (SLM): A review. *Int. J. Adv. Manuf. Technol.* **2021**, *113*, 2413–2435. [CrossRef]
15. Gockel, J.; Sheridan, L.; Koerper, B.; Whip, B. The Influence of Additive Manufacturing Processing Parameters on Surface Roughness and Fatigue Life. *Int. J. Fatigue* **2019**, *124*, 380–384. [CrossRef]
16. Zhao, Z.Y.; Wang, J.B.; Du, W.B.; Bai, P.K.; Wu, X.Y. Numerical simulation and experimental study of the 7075 aluminum alloy during selective laser melting. *Opt. Laser Technol.* **2023**, *163*, 109814. [CrossRef]
17. Wang, L.F.; Jiang, X.H.; Zhu, Y.H.; Zhu, X.G.; Sun, J.; Yan, B. An approach to predict the residual stress and distortion during the selective laser melting of AlSi10Mg parts. *Int. J. Adv. Manuf. Techol.* **2018**, *97*, 3535–3546. [CrossRef]
18. Gan, M.J.; Wu, Q.; Long, L.C. Prediction of Residual Deformation and Stress of Laser Powder Bed Fusion Manufactured Ti-6Al-4V Lattice Structures Based on Inherent Strain Method. *Mater. Res-Ibero-Am. J.* **2023**, *26*, e20220516. [CrossRef]
19. Jiang, X.H.; Ye, T.; Zhu, Y.H. Effect of process parameters on residual stress in selective laser melting of AlSi10Mg. *Mater. Sci. Technol.* **2019**, *69*, 342–352. [CrossRef]
20. Fritsch, T.; Sprengel, M.; Evans, A.; Farahbod-Sternahl, L.; Saliwan-Neumann, R.; Hofmann, M.; Bruno, G. On the determination of residual stresses in additively manufactured lattice structures. *J. Appl. Crystallogr.* **2021**, *54*, 228–236. [CrossRef]
21. Luo, C.; Qiu, J.H.; Yan, Y.G.; Yang, J.H.; Uher, C.; Tang, X.F. Finite element analysis of temperature and stress fields during the selective laser melting process of thermoelectric SnTe. *J. Mater. Process. Tech.* **2018**, *261*, 74–85. [CrossRef]
22. Shibata, K.; Kovács, A.; Kiselev, N.S.; Kanazawa, N.; Dunin-Borkowski, R.E.; Tokura, Y. Temperature and magnetic field dependence of the internal and lattice structures of skyrmions by off-axis electron holography. *Phys. Rev. Lett.* **2017**, *118*, 087202. [CrossRef] [PubMed]
23. Hu, B.; Cai, G.S.; Deng, J.L.; Peng, K.; Wang, B.X. Study on Thermodynamic Behavior, Microstructure and Mechanical Properties of Thin-Walled Parts by Selective Laser Melting. *Met. Mater. Int.* **2024**, *3*, 194–210. [CrossRef]
24. Jamnikar, N.; Liu, S.; Brice, C.; Zhang, X.L. Comprehensive molten pool condition-process relations modeling using CNN for wire-feed laser additive manufacturing. *J. Manuf. Process.* **2023**, *98*, 42–53. [CrossRef]
25. Ge, J.G.; Huang, J.; Lei, Y.P.; O'Reilly, P.; Ahmed, M.; Zhang, C.; Yan, X.C. Microstructural features and compressive properties of SLM Ti6Al4V lattice structures. *Surf. Coat. Technol.* **2020**, *403*, 126419. [CrossRef]
26. Chen, D.J.; Li, G.; Wang, P.; Zeng, Z.Q.; Tang, Y.H. Numerical simulation of melt pool size and flow evolution for laser powder bed fusion of powder grade Ti6Al4V. *Finite Elem. Anal. Des.* **2023**, *223*, 103971. [CrossRef]
27. Zhao, M.Z.; Wei, H.L.; Mao, Y.M.; Zhang, C.D.; Liu, T.T.; Liao, W.H. Predictions of Additive Manufacturing Process Parameters and Molten Pool Dimensions with a Physics-Informed Deep Learning Model. *Engineering* **2023**, *23*, 181–195. [CrossRef]
28. Yu, G.Q.; Gux, G.Q.; Dai, D.H.; Xia, M.J.; Ma, C.L.; Chang, K. Influence of processing parameters on laser penetration depth and melting/re-melting densification during selective laser melting of aluminum alloy. *Appl. Phys. A-Mater.* **2016**, *122*, 891. [CrossRef]

materials MDPI

Article

Study on Microstructure and Thermal Cracking Sensitivity of Deposited Ti6Al4V/Inconel 718 Composites Made by Two-Wire Arc Additive Manufacturing by Current

Peng Xia, Xin Ye * and Guangshun Zhang

School of Materials Science and Engineering, Shanghai University of Engineering Science, Shanghai 201620, China; xia159112@163.com (P.X.); zhang16140225@163.com (G.Z.)
* Correspondence: yexin@sues.edu.cn; Tel.: +86-67791474

Abstract: Ti6Al4V/Inconel 718 composites were prepared using arc additive manufacturing technology at different deposition currents. The properties of the composites directly influence the performance of the gradient materials, while heat input further affects the composites' properties. The results indicate that at a deposition current of 35 A, Ti elements diffuse into the Inconel 718 alloy. Increasing the current leads to the formation of brittle intermetallic compounds such as TiNi, Cr_2Ti, and Fe_2Ti in the deposited layer. At deposition currents below 50 A, no cracks appear, but cracks develop at a current of 50 A. Additionally, the microhardness of the deposited layer increases with higher deposition currents. Compared to the 35 A condition, microhardness rises by 31.51% at a current of 50 A. This research can expand the application field of the arc additive manufacturing of direct deposition Ti6Al4V/Inconel 718 composites.

Keywords: arc additive manufacturing; microstructure; Ti6Al4V/Inconel 718 composites; mechanical properties

Citation: Xia, P.; Ye, X.; Zhang, G. Study on Microstructure and Thermal Cracking Sensitivity of Deposited Ti6Al4V/Inconel 718 Composites Made by Two-Wire Arc Additive Manufacturing by Current. *Materials* **2024**, *17*, 5989. https://doi.org/10.3390/ma17235989

Academic Editors: Nikolaos Kladovasilakis, Konstantinos Tsongas and Dimitrios Tzetzis

Received: 22 October 2024
Revised: 19 November 2024
Accepted: 2 December 2024
Published: 6 December 2024

1. Introduction

The ongoing advancement of the aerospace sector has led to an increased demand for more sophisticated spacecraft, particularly in the area of dissimilar metal connection structures. These structures must now meet the dual requirements of high performance and lightweight design [1–3]. Therefore, more and more researchers have begun to pay attention to the connection of dissimilar metals and obtained Ti/Al, Mg/Al, Fe/Al, Ti/Ni, and other composite components [4–8]. Among them, Ti6Al4V alloy and Inconel 718 alloy are widely used in the aerospace field due to their excellent properties, so there is increasing attention on the connection between titanium-based alloy and nickel-based superalloy [9,10]. The structure composed of the connection of Ti6Al4V alloy and Inconel 718 superalloy can effectively reduce the mass of the parts and adapt to high-temperature environments [11,12]. However, due to the differences in thermal physical property parameters such as the thermal expansion coefficient and thermal conductivity coefficient of dissimilar metals, cracks will occur in the connection process of titan-based alloy and nickel-based superalloy [13–15], and the preparation of gradient materials can improve this situation. In addition, when dissimilar metals are connected, composite materials are often formed at the joint, in which brittle intermetallic compounds (IMCs) such as Ti_2Ni and Ni_3Ti are easily formed, which further increases the brittleness of dissimilar joints [14,16,17]. Therefore, the microstructure and mechanical properties of composite materials play an important role in the preparation of gradient materials.

In the process of dissimilar metal connection, new composite materials will inevitably be formed, so some researchers achieve a dissimilar metal connection by changing the alloy's composition layer by layer. Sun et al. [18] prepared the Inconel 625/Ti6Al4V component using laser additive manufacturing technology and studied the changes in the composition and structure of the alloy from Ti6Al4V to Inconel 625. The results show that the phase changes to α+β→α+β+Ti_2Ni→Ti_2Ni+β→Ti_2Ni+Cr_2Ni+γ-Ni, and a micro-crack occurs in the diffusion zone at 800 °C. Meng et al. [19] prepared Inconel 625/Ti6Al4V functionally graded materials utilizing laser synchronous preheating/non-preheating and found that preheating had a great influence on the formation of cracks and precipitates. Song et al. [10] used laser additive manufacturing technology to prepare composite materials with different Ti contents. The results showed that with the increase in the Ti content, the main phase of the prepared component changed to γ→Fe_2Ti→TiNi. Onuike et al. [20] used laser additive manufacturing technology to prepare Inconel/Ti6Al4V gradient material and introduced a new phase Cr_3C_2 by introducing VC as the intermediate layer to avoid the generation of a brittle phase. Wang et al. [15] prepared the Ti6Al4V/Inconel718 component using linear laser additive manufacturing technology. The results showed that the phase microstructure of the Ti6Al4V/Inconel 718 component was γ+Laves→α+β+Ti_2Ni+TiNi+Ni_3Ti→α+β from the bottom to the top. However, when the welding wire is used as a raw material for laser additive manufacturing, there will be stratification. Most researchers prefer to use lasers as heat sources to prepare functionally graded materials. However, because the laser is used as the deposition heat source, it needs high power to melt the wire or powder, which greatly increases the deposition cost. When the arc is used as the deposition heat source, it is not necessary to use high power to melt the wire material, and arc additive manufacturing has the advantages of high product efficiency, a high material utilization rate, and low equipment cost [21]. Huang et al. [21] used arc additive manufacturing technology to prepare TiNi gradient materials by controlling the wire feed speeds of Ti6Al4V and Inconel 625. The results showed that the use of arc could make the phase distribution of Ti_2Ni and TiNi uniform and improve the compression performance and surface wear performance of TiNi components. Lu et al. [22] used arc additive manufacturing technology to prepare the functional gradient of TA1/Inconel 625 and studied its microstructure and mechanical properties. However, only a few studies have preliminarily discussed the use of arc additive manufacturing technology to connect Inconel 718 and Ti6Al4V alloys. In summary, there are few studies on the preparation of Ti6Al4V/Inconel 718 gradient materials by arc additive manufacturing; in particular, the discussion on microstructure and mechanical properties is relatively limited.

The premise of preparing gradient materials is to prepare intact composite materials, so the microstructure and mechanical properties of composite materials often affect the properties of gradient materials. In the process of deposition, heat input will affect the microstructure and mechanical properties of composite materials. In this paper, Ti6Al4V and Inconel 718 wires are selected as Ti6Al4V and Inconel 718 welding equipment, and the wire feeding speed of the two wires is fixed to prepare Ti6Al4V/Inconel 718 composite materials with different deposition currents. The effects of different deposition currents on the microstructure and mechanical properties of Ti6Al4V/Inconel 718 composites were investigated. By comparing the morphologies of the intermetallic phases in the composites under different deposition currents, the causes of the crack sections were analyzed. In addition, the microhardness of the samples prepared by different deposition currents was evaluated. It can expand the application field of arc additive manufacturing direct deposition Ti6Al4V/Inconel 718 composites.

2. Experimental Details

The experiment adopts a single arc deposition system, which mainly includes a TIG welding machine, servo motor mobile platform, and wire feeding mechanism. The working principle is shown in Figure 1. The Ti6Al4V and Inconel 718 wires are heated by an arc heat source. The two wires are placed on one side and the wire feeder can achieve different wire feed speeds. The system uses 99.99% high-purity argon as the protection gas.

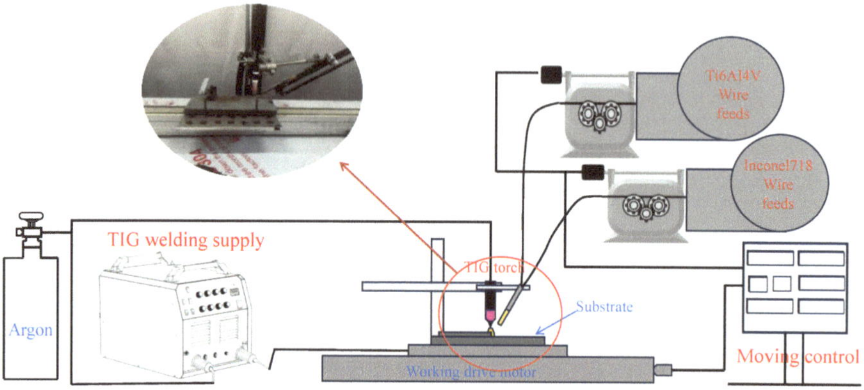

Figure 1. D-WAAM device diagram.

The substrate was Inconel 718 and its dimensions were 100 mm × 50 mm × 4 mm. Before the experiment, the surface was polished with an angle grinder, the oxide layer was removed, and the surface was washed and dried with acetone and alcohol. A layer of Inconel 718 was deposited before the Ti6Al4V/Inconel 718 composite layer was deposited. Continued deposition occurred on the deposited Inconel 718 layer. The diameter of the Ti6Al4V and Inconel 718 wires is 1.2 mm, and the chemical composition of the two wires is shown in Tables 1 and 2. The composition of the alloy is controlled by considering the wire feed speed (V_i), density, and diameter [23]. Table 3 shows the deposition parameters. The mass fraction (W_x) and atomic fraction (A_x) can be calculated using the following formula [24]:

$$W_x = \frac{\sum V_i d_i \rho_i W_{x_i}}{\sum V_i d_i \rho_i} \quad (1)$$

$$A_x = \frac{W_x / M_x}{\sum W_x / M_x} \quad (2)$$

where i is a metal wire (i = 1, 2, 1 means Ti6Al4V, 2 means Inconel 718), W_{x_i} (x is an element in the Ti6Al4V and Inconel 718 alloys; i = 1, 2) is the mass fraction of an element in the wire, $V_i(i = 1, 2)$ is the wire feeding speed of the wire, and the unit is mm/s. $d_i(i = 1, 2)$ is the diameter of the wire in mm. M_x (x is an element in the Ti6Al4V and Inconel 718 alloys) is the relative atomic mass and $\rho_i(i = 1, 2)$ is the density of the wire, expressed in g/cm^3.

Table 1. Chemical composition of Ti6Al4V welding wire (wt%).

Al	V	Fe	C	N	H	O	Ti
5.9	3.97	0.05	0.10	0.05	0.01	0.10	Bal.

Table 2. Chemical composition of Inconel 718 wire (wt%).

Ni	Cr	Nb	Mo	Ti	Al	Si	Fe
53.61	18.76	4.86	3.12	1.05	0.49	0.18	Bal.

The sample is cut along the longitudinal section, and the small sample is polished with 400–2000 purpose SiC sandpaper. Further polishing was carried out with a 1.5 particle size suspension. Finally, Kroll reagent (HF: HNO_3: C_2H_5OH=7:2:1) was used for etching for about 10 S. The phase structure morphology between dissimilar metals was observed by a laser confocal microscope, and then the phase structure was analyzed by a high-magnification electron microscope. The microstructure of the samples was analyzed by scanning electron microscopy (SEM) and energy dispersive spectroscopy (EDS). The phase composition of the sample was analyzed by X-ray diffraction (XRD). The microhardness of the composite was measured by the Hv-1000 Vickers microhardness tester. The loading force was 200 g, and the residence time was 15 s.

Table 3. Deposition parameters.

Parameter	Value
Wire feed rate (Ti6Al4V)	400 mm/min
Wire feed rate (Inconel 718)	800 mm/min
Gas flow	15 L/min
Plate speed	0.9 mm/s

3. Results and Discussion

3.1. Single-Layer Deposition Analysis

To determine the appropriate process parameters, the ratio of two types of welding wire is adjusted by varying the speed of the wire feeder. After fixing the wire feeding speed for both types of welding wire, different deposition currents are applied to prepare the sediment. Figure 2 shows the deposition morphologies under different deposition currents. As the current increases, the roughness of the deposited layer also increases significantly, primarily due to the increase in the viscosity of the liquid metal. Van et al. [25,26] studied the relationship between heat input and forming quality, and the results showed that the lower the heat input, the smoother the surface of the sedimentary components. This is consistent with the morphology of the samples deposited in this work. At higher cooling rates, the wettability and diffusivity of liquid metals are reduced, resulting in poorer adhesion. Furthermore, when the current reaches 50 A, cracks appear on the surface of the layer.

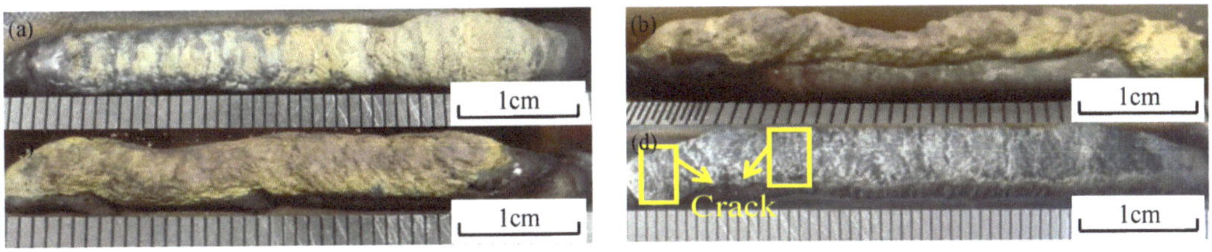

Figure 2. Sedimentary morphologies of different sedimentary currents: (**a**) 35 A, (**b**) 40 A, (**c**) 45 A, and (**d**) 50 A.

3.2. XRD Phase Analysis

The wire feed speed of Inconel 718 is twice that of Ti6Al4V. According to Formulas (1) and (2), when the element ratio of titanium to nickel is 1:2, a deposited layer is prepared using these two welding wires. During the deposition process, Ti-Ni intermetallic compounds are formed. In addition, Inconel 718 alloy also contains a certain amount of Cr and Fe, which may also form intermetallic compounds with Ti. The X-ray diffraction method was used to analyze the phases of the sediments prepared under different sedimentary currents, as shown in Figure 3. All the samples exhibited the TiNi phase, Ni_3Ti phase, Cr_2Ti phase, α-Ti phase, and Fe_2Ti phase.

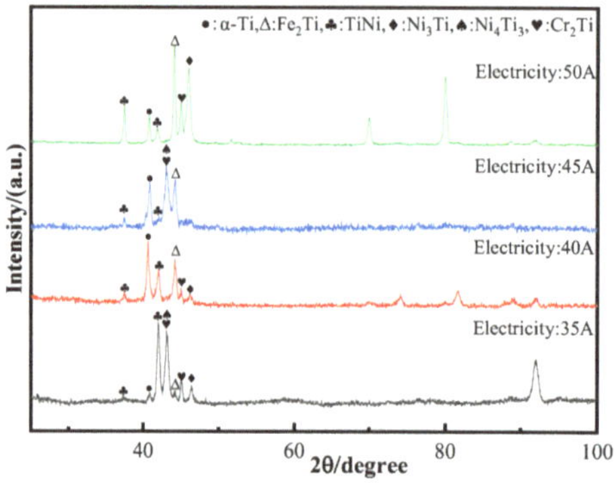

Figure 3. X-ray diffraction spectra of deposited products under different deposition currents.

3.3. Organizational Characterization by Optical Microscopy

To study the effect of different heat inputs on the microstructure of the same composite material ($V_{Inconel\ 718}$: $V_{Ti6Al4V}$=2:1), the metallographic morphology of the samples was observed under different deposition currents using an optical microscope. Figure 4 displays the optical microscopy characterization for each deposition current. Figure 4a shows the optical microstructure at a deposition current of 35 A, with the Ti6Al4V/Inconel 718 composite in the upper part and the Inconel 718 layer in the lower part. The upper section is predominantly composed of columnar crystals, intermingled with some long black bars among the branches of the columnar crystals. In the WAAM process, the solid–liquid interface gradually advances into the liquid phase region. During this advancement, the solute undergoes redistribution, resulting in a phenomenon of component supercooling within a certain distance. This instability of the flat solid–liquid interface ultimately leads to the formation of columnar crystals. As the liquid phase region is influenced by solute redistribution, solute aggregation occurs, which is manifested as microscopic segregation [27]. From the Figure 4a, it can be observed that tiny secondary dendrites are present on the sides of the columnar crystals, with an average distance of 5.16 μm between these secondary dendrites.

With the increase in the deposition current, the columnar crystals in the deposited layer gradually undergo differentiation. Figure 4b shows the optical microstructure at a deposition current of 40 A. From the Figure 4b, it can be observed that some of the grains are arranged in an orderly manner, while a discontinuous black phase is present between the grains. In addition, long straight columnar crystals can be observed in the Figure 4b, which is formed due to the instability of the solid–liquid interface resulting from the supercooling of the composition. Under conditions of rapid cooling, columnar crystals are formed more readily. In comparison to Figure 4a, when the deposition current is 40 A, the heat input increases, leading to a longer temperature residence time on the workpiece under the same heat dissipation conditions. As a result, the cooling rate becomes relatively slower. At the microscopic level, the number of columnar crystals decreases, with some larger columnar crystals gradually transforming into smaller grains, leading to an increase in the spacing between columnar crystals. As some of the heat is conducted in the direction of the deposited additive, a negative temperature gradient develops in the liquid phase, perpendicular to the crystal axis. This results in the formation of a secondary crystal axis along the primary crystal axis [27]. The presence of secondary dendrites can be observed in the Figure 4b, with the average distance between the secondary dendrites measuring 6.38 μm.

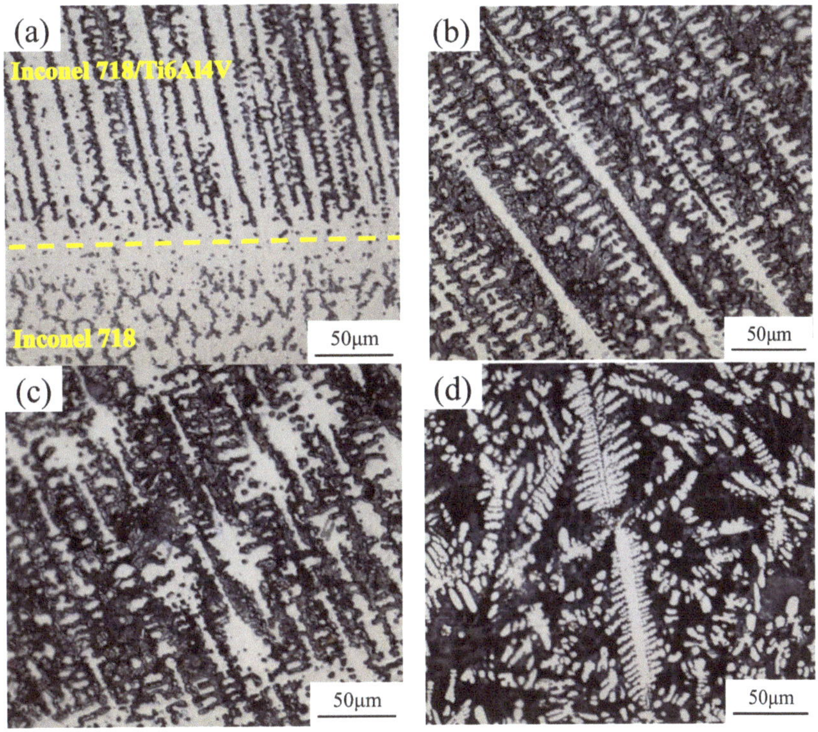

Figure 4. Optical microscope characterization of samples under different deposition currents: (**a**) 35 A, (**b**) 40 A, (**c**) 45 A, and (**d**) 50 A.

When the deposition current is further increased, the metallographic morphology of the deposited layer is shown in Figure 4c. This Figure 4c illustrates the optical microstructure at a deposition current of 45 A, where fine columnar crystals and "leaf-like" grains can be observed. As the current increases, the grains grow further, with some secondary dendrites fusing during the growth process to form larger "leaf-like" grains that merge with the columnar crystal axis.

Figure 4d illustrates the optical microstructure at a deposition current of 50 A, where typical dendritic columnar crystals are observed, and the content of epitaxial crystals has increased. In comparison to the deposition current of 45 A, the increase in the heat input slightly extends the heat dissipation time, allowing for further grain growth and an increase in secondary dendrites. The average distance between the secondary dendrites is 3.96 μm. Additionally, a significant presence of grayish-black intra-crystalline structures can be observed in this region, which will necessitate further examination through EDS scanning analysis.

Figure 5 presents a schematic diagram illustrating the deposition process and the grain evolution of the sedimentary sample as the current increases. As the deposition current is gradually increased, the columnar crystals in the deposited samples evolve from a "long strip" morphology to a "rod-like" structure. With further increases in current, these "rod-like" dendrites begin to merge, resulting in the formation of "leaf-like" dendrites. When the current reaches 50 A, the "leaf" dendrites split into smaller dendrites. The sampling method for the sedimentary sample involves examining the cross-section of the additive body. When the deposition current is low (<50 A), the growth direction of the columnar crystals is perpendicular to the direction of the additive. However, as the deposition current reaches 50 A, the growth direction of a few columnar crystals remains perpendicular to the additive body, while most columnar crystals become parallel to the additive. This change may be attributed to the increased deposition current, which leads to an increase in heat input. In the Wire Arc Additive Manufacturing (WAAM) process, excessive heat diffuses along the formed additive, causing the growth direction of the columnar crystals to favor a parallel orientation with the additive body.

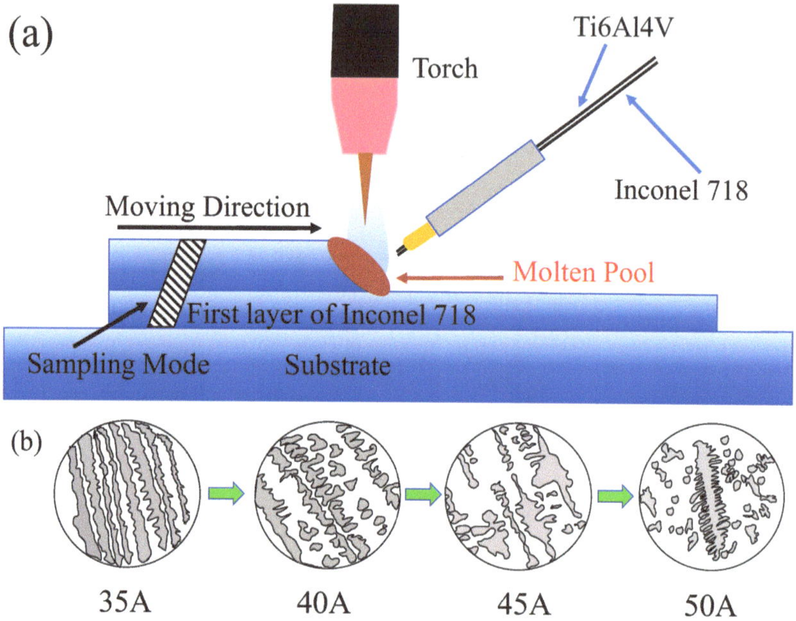

Figure 5. (**a**) Deposition diagram and (**b**) grain evolution diagram of the current 35 A–50 A deposition sample.

Figure 6 presents the statistical results of secondary dendrite arm spacing (SDAS) at different deposition currents. At deposition currents of 40 A and 45 A, the SDAS values are similar and larger than those at 35 A and 50 A. Notably, the minimum SDAS at 50 A is recorded at 3.96 μm. As the current increases, the heat input also rises, leading to an increase in the deposition temperature. This change enhances atomic mobility, increases the kinetic energy of certain atoms, and reduces the interfacial energy among grains, thereby promoting grain growth. The grain growth process is also clearly illustrated in Figure 5. When the deposition current is lower than 50 A, the change in secondary dendrite spacing SDAS is minimal, and all measured values are lower than those observed at 50 A. This phenomenon may be attributed to the limited heat input in the sedimentary layer and an excessively high cooling rate, which hinders the kinetic energy required for grain growth. Consequently, atomic migration is impeded, leading to a cessation of grain growth, and the secondary dendrite spacing does not exhibit significant changes. When the deposition current reaches 50 A, the increased heat input facilitates the further growth of some grains, while additional secondary dendrites also form, resulting in a decrease in secondary dendrite spacing. The morphology of the grains gradually shifts from columnar to dendritic, with the thickness of the columnar crystals also decreasing.

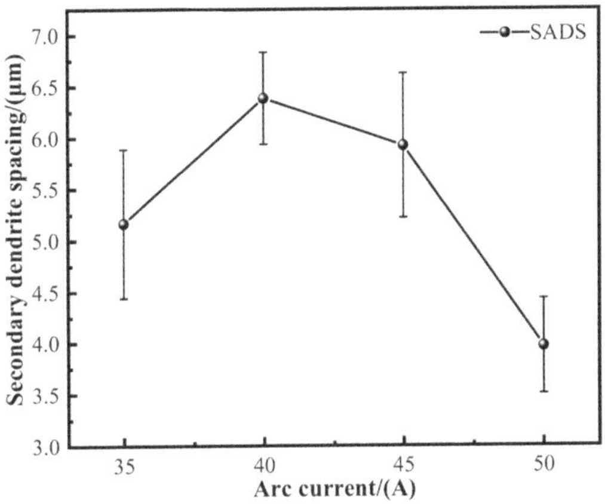

Figure 6. Statistical results of secondary dendrite spacing for each sedimentary current.

3.4. SEM and Element Content Analysis

To investigate the tissue composition of samples deposited under various current conditions, scanning electron microscopy (SEM) and energy dispersive spectrometry (EDS) point analyses were conducted on each deposited layer. The detailed results are presented in Figure 7 and Table 4. By combining the information from Figure 4a, Figure 7a,e, and Table 4, it can be concluded that the black phase (P1) between the columnar crystals exhibits the highest carbon (C) content, suggesting the presence of MC carbide. MC carbides play a crucial role in nickel-based alloys. Even a small amount of MC carbides can enhance strength by inhibiting dislocation movement and stabilizing grain boundaries [15]. At the position corresponding to the crystal axis (P2), the concentrations of Ni, Cr, and Fe are elevated. This may be attributed to the low deposition current in the WAAM process, which leads to a rapid cooling rate and results in an uneven distribution of Ti and Ni in the molten pool. Therefore, Ni and Ti do not fully integrate, with some Ti diffusing into the matrix and forming a eutectic structure. Consequently, when the deposition current is set at 35 A, a small amount of the TiNi phase is formed. In the precipitated phase (P3), the Ti content is 68.52 at%, while the Ni content is 23.47 at%. This suggests that the precipitated phase may consist of Ti_2Ni particles.

As the electric current increases, the columnar crystal structure in the deposited layer gradually differentiates, as illustrated in Figure 7b. Figure 7f provides a localized magnification, revealing numerous small "villous" structures at the grain boundaries (P5). According to energy dispersive spectroscopy (EDS) scanning analysis, the concentrations of Ti, Ni, Cr, and Fe do not show significant differences. This phenomenon may be attributed to the increased heat input, allowing the Ti in the molten pool to fully diffuse and form intermetallic compounds with Ni, Cr, and Fe. Thus, it can be inferred that this "fluff"-like phase may consist of a eutectic structure composed of Ti, Ni, Cr, and Fe elements. At the P4 position of the grain (Figure 7f), the Ti element content is significantly higher than the Ti element content at the P2 position (Figure 7e). This may be attributed to the increased heat input associated with the higher current, which facilitates the binding of Ti atoms to Ni atoms. Consequently, the increased heat input promotes atomic diffusion and enhances the formation of compounds with other atoms. The concentrations of Ti, Cr, and Fe atoms in both the dendrite axis and interdendritic structure are similar. By correlating this with the X-ray diffraction (XRD) data, it can be inferred that Ti atoms interact with Cr, Fe, and Ni atoms to form intermetallic compounds such as Cr_2Ti, Fe_2Ti, and TiNi. Additionally, Nb and Mo atoms are not detected in the grains (P4). This may be attributed to the diffusion of heavy elements, such as Nb and Mo, into the interdendritic region during solidification, leading to their separation from the eutectic reaction with the remaining liquid phase between the dendrites [15].

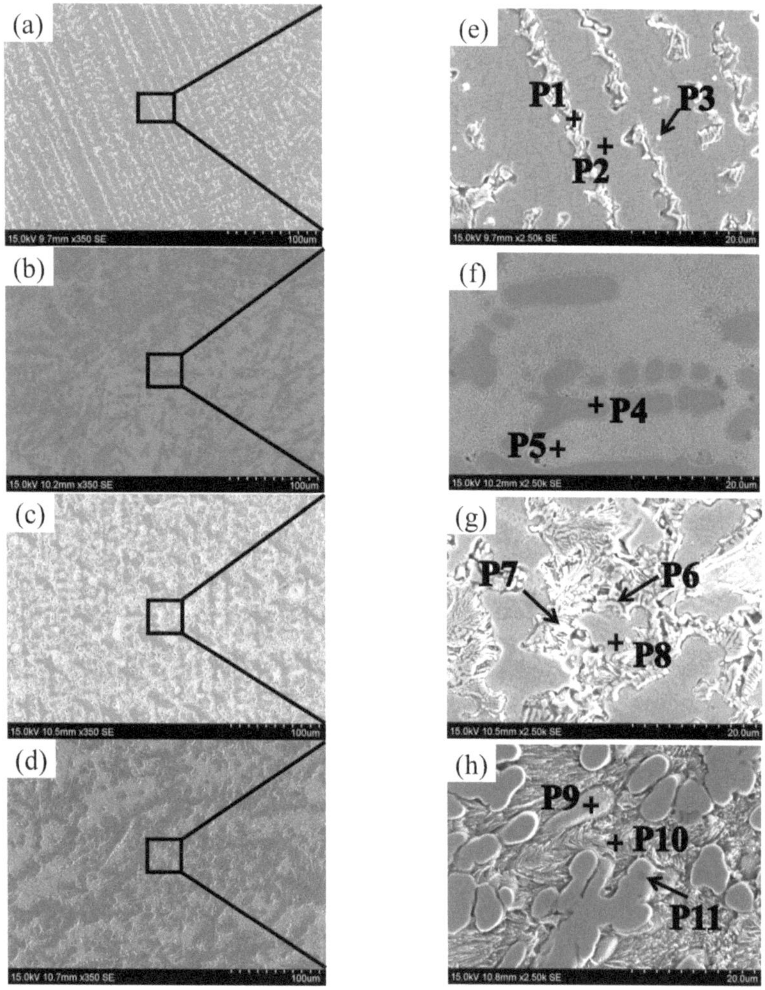

Figure 7. SEM scanning topography and EDS measuring point locations: (**a**) 30 A, (**b**) 40 A, (**c**) 45 A, (**d**) 50 A, (**e**) partial magnification in a, (**f**) partial magnification in b, (**g**) partial magnification in c, and (**h**) partial magnification in d.

Figure 7c illustrates the microstructure of the sample sediment obtained with a deposition current of 45 A, while Figure 7g provides a local magnification. From Figure 7g, it is evident that numerous needle-like precipitates are present at the grain boundaries. After conducting energy dispersive spectrum (EDS) scanning analysis, it was observed that the ratio of Ni atoms to Ti atoms in the needle-like structure is approximately 3:1. According to the Ti-Ni phase diagram (Figure 12a), it can be inferred that the needle-like structure may be Ni_3Ti, while the grain structure likely consists of γ + TiNi. In the precipitates located at position P7, the highest concentration of carbon C was detected, suggesting that the microstructure of this particular phase is MC carbide.

Table 4. EDS point scanning results of each micro-area (at.%).

Area	Ti	Ni	Cr	Fe	Nb	Mo	C
P1	4.05	17.78	12.79	10.54	6.55	2.52	44.45
P2	6.1	48.14	22.89	19.64	-	-	-
P3	68.52	23.47	-	-	-	-	-
P4	17.15	43.48	21.31	16.02	-	-	-
P5	20.22	27.59	20.96	20.86	6.10	3.16	-
P6	12.87	33.63	21.89	19.56	6.18	3.66	-
P7	5.23	11.72	12.91	11.01	4.57	2.13	51.63
P8	8.20	44.65	23.81	21.59	23.81	-	-
P9	22.48	26.71	21.91	21.17	4.12	2.59	-
P10	22.94	45	16.96	15.10	-	-	-
P11	12.60	17.08	14.59	13.68	3.71	2.10	35.58

Figure 7d illustrates the microstructure of the sample sediment obtained with a deposition current of 50 A, while Figure 7h provides a local magnification. From Figure 7h, it can be observed that numerous acicular structures are present between the dendrites. According to the EDS scanning results, the atomic contents of Ti and Ni in the intergranular acicular structure are 22.94 at% and 45 at%, respectively. The atomic ratio of Ni to Ti is approximately 2:1. Together with the X-ray diffraction analysis results, it can be concluded that the needle structure is likely a nickel-rich TiNi phase. At the position of the precipitated phase P9, the atomic ratio of Ti to Ni is approximately 1:1, and the atomic contents of Cr and Fe are similar. It can be inferred that the precipitated phase may be a eutectic structure composed of Ti, Ni, Cr, and Fe elements, potentially forming brittle intermetallic compounds such as TiNi, Cr_2Ti, and Fe_2Ti. At position P11 in Figure 7h, a black precipitated phase can be observed on the dendrite axis. According to the EDS scanning results, the carbon content at P11 is higher than that of the other elements. This may be attributed to excessive heat input, leading to the partial diffusion of MC carbide into the grain during solidification. It has been reported that a small amount of MC carbide can inhibit dislocation movement and stabilize grain boundaries, thereby strengthening the material. However, an excessive amount of MC carbide can cause stress concentration and reduce the durability of the alloy [15]. When the deposition current reaches 50 A, cracks begin to appear in the sedimentary layer. This may be attributed to the stress concentration caused by the accumulation of MC carbides, as shown in Figure 2d. A detailed analysis of the crack section will be provided in the following section.

Based on the analysis above, heat input significantly affects the element content between the dendrite axis and the dendrite. Furthermore, under the same deposition current, there are considerable differences in the element content between the dendrite axis and the dendrite, indicating that the segregation mechanism is intergranular segregation. The XRD results did not detect Ti_2Ni intermetallic compounds, possibly due to their low content. According to the Ti-Ni binary phase diagram (Figure 12a), the formation of Ti_2Ni requires the Ti content to exceed that of Ni. Additionally, since the volume of Inconel 718 alloy in this experiment is twice that of Ti6Al4V alloy, the formation of Ti_2Ni is unlikely. Generally, when the two alloy wires are melted in the melt pool, the internal elements should ideally diffuse evenly. However, during the arc additive manufacturing process, the extremely fast cooling rate and small deposition current lead to solidification before thorough diffusion occurs. This results in significant variations in the element content in certain areas. Although this issue improves with a gradual increase in the deposition current, cracks may still develop in the deposited layer when the current reaches a certain level due to excessive heat input.

According to SEM-EDS analysis, the Ni content along the dendrite axis is higher than the Ti content under various deposition currents, while the Ti content along the dendrite axis increases with the rising deposition current. Figure 8 illustrates the proportion of the dendrite axis. At a deposition current of 35 A, the dendrite axis proportion reaches a maximum of 80.36%. However, this proportion decreases overall as the deposition current increases. A comparison of the element content in the interdimeric and dendritic regions reveals that, with an increasing deposition current, the alloying element concentrations in these two areas become more similar, indicating a trend toward a more uniform phase distribution.

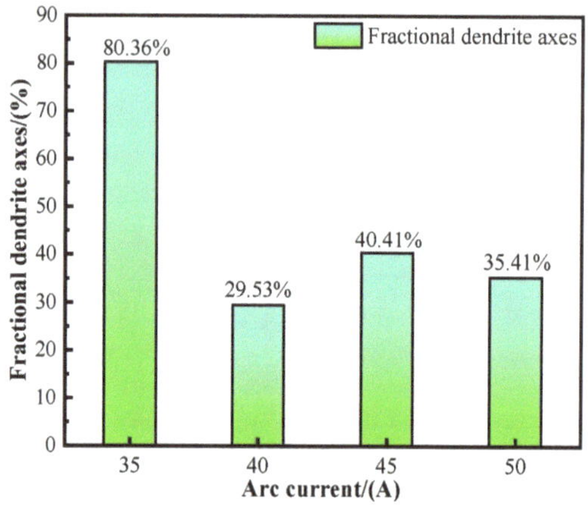

Figure 8. Proportion of dendrite axes.

3.5. *Cracks*

Titanium alloys are important structural materials, while nickel-based alloys are considered superalloys. Due to their high thermal expansion coefficients, cracks are prone to develop during deposition. In the WAAM process, no cracks were observed in the deposited layers at deposition currents below 50 A. However, at a deposition current of 50 A, cracks were detected in the deposited layer. Figure 9 illustrates the macroscopic view of the partial cracks, as well as high-rate SEM images. No cracks were found in the first layer, but the crack section in the second layer Ti6Al4V/Inconel 718 appeared disconnected, with an uneven crack morphology. High-magnification SEM images (shown in Figure 9b–d) of the crack section reveal a microfracture morphology characterized by river and cleavage patterns, indicating a cleavage fracture mechanism.

To investigate the phase composition of the crack section, an EDS scanning analysis of its microstructure was conducted. Figure 10 presents a high-magnification SEM image of the fracture bottom, revealing a precipitated phase A. EDS analysis indicates that the atomic ratio of oxygen to iron is approximately 1.6:1, suggesting that the precipitate is likely iron oxide. At position B, the ratio of nickel to titanium atoms is about 1:1, with the chromium and iron content being similar to that of titanium and nickel. However, the carbon content is relatively high, leading to the speculation that the main phase at position B is MC carbide. Previous studies have shown that small amounts of MC carbide can impede dislocation movement and stabilize grain boundaries, thus providing strengthening; however, excessive MC can result in stress concentration, reducing the alloy's durability and contributing to crack formation [15]. At position C, the nickel atom content is highest, with titanium, chromium, and iron content being comparable. This leads to the speculation that the phase at this position may be a eutectic structure of brittle intermetallic compounds formed by Ni, Ti, Cr, and Fe.

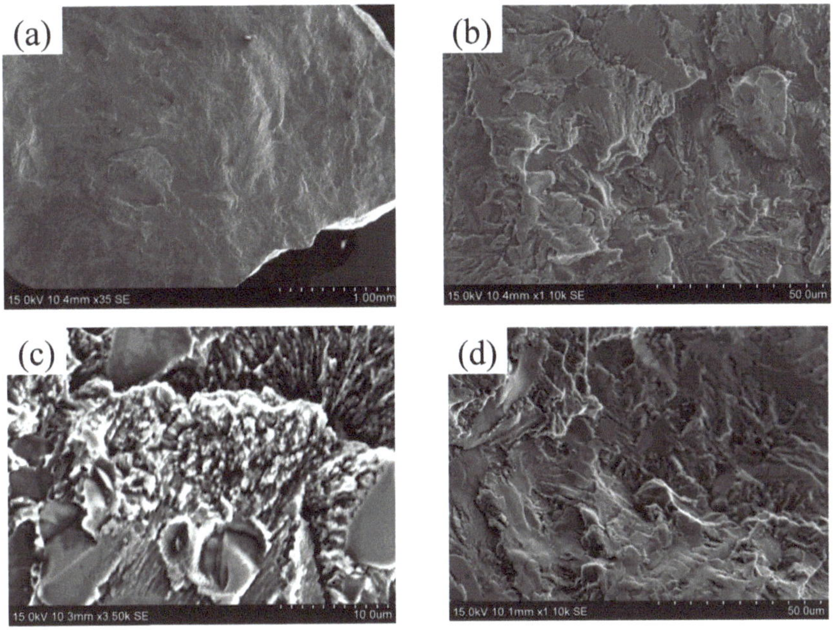

Figure 9. SEM images at a deposition current of 50 amps: (**a**) macroscopic profile of some cracks, (**b–d**) high magnification SEM images.

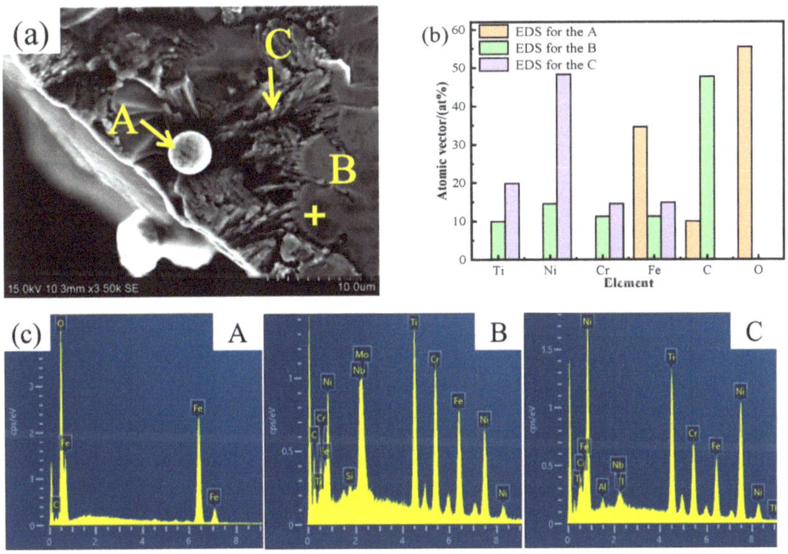

Figure 10. (**a**) High-magnification SEM image of the bottom of the fracture at a deposition current of 50 A. (**b**) Comparison of elemental content, (**c**) EDS-measured content map at three points in (**a**).

As can be seen from Figure 9, the cross-section is distributed with a large number of river patterns and a "step" shape. According to XRD results and EDS scanning analysis, brittle intermetallic compounds of Ti-Fe, Ti-Ni, and Ti-Cr series are distributed in the sedimentary layer prepared by WAAM [28,29]. When the deposition current is below 50 A, the heat input is minimal, and the yield strength of the Ti6Al4V and Inconel 718 alloys exceeds the thermal stress in the deposited layer, resulting in no crack formation. However, when the deposition current exceeds 50 A, the stress concentration in the deposited layer surpasses the yield strength of the alloy, leading to crack formation.

3.6. Mechanical Properties

Figure 11 presents the microhardness characteristics of samples subjected to various deposition currents, as measured by a microhardness tester. At a deposition current of 35 A, the average microhardness is 550.14 HV. This value increases to 658.27 HV at 40 A and reaches 659.79 HV at both 45 A and 50 A. The maximum recorded average microhardness is 723.50 HV. Previous studies report that the microhardness of Inconel 718 is approximately 250 HV [30]. The results indicate a significant enhancement in the microhardness of the Ti6Al4V composite, which increases progressively with higher deposition currents.

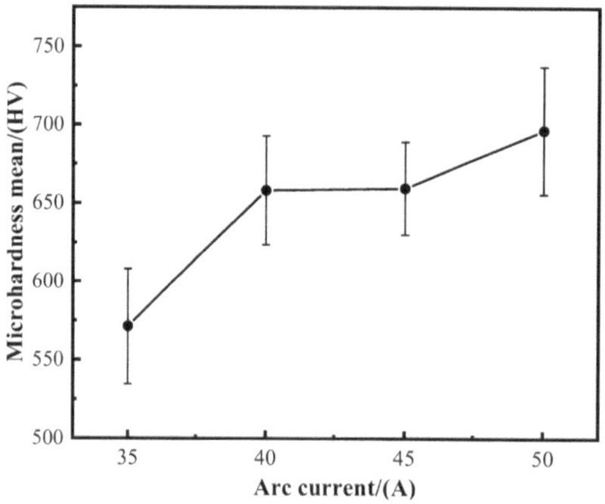

Figure 11. Microhardness characteristics of samples with different deposition currents.

As the deposition current increases, the heat input also rises, leading to changes in the secondary dendrite spacing, grain size, and morphology of the deposited layer. A higher deposition current results in smaller grains within the deposited layer, which increases the relative area of the grain boundaries and hinders dislocation movement. This obstruction contributes to an increase in yield strength, subsequently affecting the microhardness of the sample. Previous analysis shows that at a deposition current of 35 A, the deposited layer primarily consists of columnar crystals. These crystals are predominantly nickel-based alloys with a small amount of Ti distributed within them through diffusion. The incorporation of Ti elements enhances the microhardness of the material.

At a deposition current of 40 A, the microstructure of the deposited layer is primarily composed of intermetallic compounds such as TiNi, Cr_2Ti, and Fe_2Ti. Both Cr_2Ti and Fe_2Ti are hard and brittle intermetallic phases that contribute to the overall hardness of the material. When the deposition current increases to 45 A, there is little change in the microhardness compared to 40 A. This stability may be attributed to the microstructure along the dendrite axis, which is predominantly a nickel-based alloy, with a small distribution of Ti. The content of hard and brittle intermetallic compounds between the dendrite axis and the dendrites is minimal; however, MC carbides are located near the grain boundaries. These carbides hinder dislocation movement, thereby enhancing the material's hardness and potentially explaining the limited change in microhardness.

When the deposition current is 50 A, the content of hard brittle intermetallic compounds on the dendrites axis and the dendrites is relatively higher, and it can be seen from Figure 7 that MC carbides are also distributed on the dendrites axis. It has been reported that a small amount of MC carbide provides strengthening by preventing dislocation movement and stabilizing grain boundaries [15], so the corresponding hardness is relatively increased. When the current is increased to 40 A, the microhardness is increased by 19.65% compared to the current of 35 A. When the deposition current increases to 50 A, the microhardness increases by 31.51% compared with that when the deposition current is 35 A.

4. Discussion

4.1. Solidification of Composite Materials Under Different Deposition Currents

In the composite materials prepared by Ti6Al4V and Inconel 718 alloys, the intermetallic compounds are mainly dominated by the Ti-Ni binary system. Figure 12 shows the Ti-Ni binary phase diagram [31]. According to the analysis of the Ti-Ni binary phase diagram, when the content of Ti and Ni atoms is close to 1:2, the eutectic reaction will occur, and new phase TiNi and Ni_3Ti metal compounds will be formed in the composite material. However, Inconel 718 alloy contains more elements, and in the WAAM process, elements other than Ni in Inconel 718 alloy will still combine with Ti elements to form Ti-Fe phase and Ti-Cr phase. It has been reported that Fe_2Ti metal compounds are described as the C14_LAVES phase [32] in crystallography, while Cr_2Ti metal compounds are also described as the Laves phase [33].

The theoretical basis of the thermodynamic calculation is the CALPHAD method (computational phase diagram) [34]. When the Gibbs energy of the whole system is at its minimum, the system is in equilibrium. In some regions, the equilibrium state may be a mixture of various phases or simply a stable single phase. In a multicomponent system, the Gibbs energy of the phase can generally be expressed by the following equation:

$$G_m = \sum_i x_i G_i^o + RT \sum_i x_i lnx_i + \sum_i \sum_j > 1 x_i x_j \sum_V \Omega_V (x_i - x_y)^2 \quad (3)$$

where G_m is the Gibbs energy of the phase, $\sum_i x_i G_i^o$ is the Gibbs energy of pure components, $RT \sum_i x_i lnx_i$ is the ideal entropy, $\sum_i \sum_j > 1 x_i x_j \sum_V \Omega_V (x_i - x_y)^2$ is a paired interaction term with interaction parameters, x_i, x_y are the molar fraction of the different alloy components, T is temperature, and R is a gas constant. Ω_V is the interaction coefficient that depends on the value of V. When the value of V provides values of 0 and 1, it corresponds to the normal solution model. In practice, the value usually does not exceed 2 [35].

The equilibrium phase diagram of multicomponent alloy and the fraction of phase equilibrium at each temperature can be calculated by the thermodynamic calculation principle [36]. Although there is a deviation between the non-equilibrium heating and cooling process and the theoretical temperature in the equilibrium state, the relative magnitude relationship between each temperature is unchanged, so the theoretical temperature in the equilibrium state can be used to analyze the solidification and precipitation of each phase in the cooling process.

In this work, EDS analysis was carried out on the dendrite axes and interdendrites in the composite materials under different deposition currents, the average alloying elements contents were obtained, and the solid–liquids theoretical temperature was calculated to obtain the precipitation temperature of the corresponding components. The results show that no Laves phase is found in the composite prepared when the deposition current is 35 A, that is, no Laves phase is precipitated during the cooling process. During the cooling process, γ, η, and BCC phases were precipitated from the dendritic axis. It has been reported that the η phase is an intermetallic compound phase based on the Ni_3Ti component [37]. The structure of BCC is body-centered cubic, and the structure of the TiNi phase is reported to be body-centered cubic [31].

Figure 12b shows the mass fraction and temperature curves of precipitation and liquefaction of γ phase, η phase, and BCC phase calculated under the equilibrium state when the deposition current is 35 A. It can be seen that the liquid phase temperature of the dendrite axis is high, so the dendrite axis solidifies first in the cooling process, and the dendrite axis will precipitate the γ phase, η phase, and BCC phase in turn during the cooling process. The solidification process involves the redistribution of solute elements. Considering that the precipitated phase is mainly related to Ni and Ti, it can be seen from the analysis of the Ti-Ni binary phase diagram in Figure 12a that the precipitated phases on the dendrite axis are γ-Ni, Ni_3Ti, and a small part of TiNi. When the temperature is higher than 1353.87 °C, all liquefaction is the liquid phase. When the temperature is lower than 1350 °C, the γ phase begins to precipitate. At this temperature, the dendrite axis begins to solidify, because the high-temperature existence time is very short, so the molten pool in the deposition process is basically in the solid–liquid mixing stage; in the cooling process, the dendrite and dendrite axis will form a liquid film, the liquid metal formation of heat absorption and surface tension mechanical action, resulting in the growth of the grain being limited and hindered, and then the austenite grain boundary belongs to the chemical grain boundary [38,39].

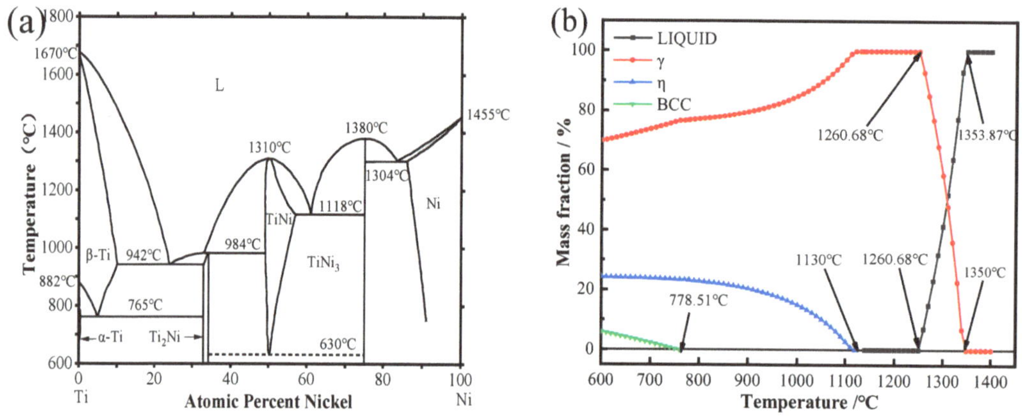

Figure 12. (**a**) Ti-Ni binary phase diagram [26] and (**b**) phase equilibrium fraction diagram of the composite when the deposition current is 35 A.

Figure 13 shows the equilibrium fraction diagram of the composite material when the deposition current is 40 A, and the calculated results show that the temperature of the dendrite axis (Figure 7f-P4) and interdendrite (Figure 7f-P5) for complete liquefaction to liquid phase is 1172.76 °C. However, the precipitation temperature of the interdendritic Laves phase is 1366.21 °C, indicating that the interdendritic Laves phase is the first to precipitate during the cooling process, and the temperature at which its content reaches the highest point is 1144.16 °C. When the temperature is reduced to around 1181.76 °C, the Ni_3Ti phase and TiNi phase on the dendritic axis begin to precipitate. During the precipitation of the Laves phase between dendrites, the Ni_3Ti phase and TiNi phase on the dendritic axis also precipitate successively, and the content of the Ni_3Ti phase hardly changes with the decrease in temperature. It is worthy of note that when the temperature is reduced to 865.82 °C, there is a discernible decrease in the TiNi phase content, accompanied by the onset of σ phase precipitation. The Laves phase has a high melting temperature, so the dendrites solidify first, which forms a liquid film with the dendrite axis, which retains the dendritic structure state, indicating that the dendrites are dispersed during the whole cooling process, as shown in Figure 7b,f.

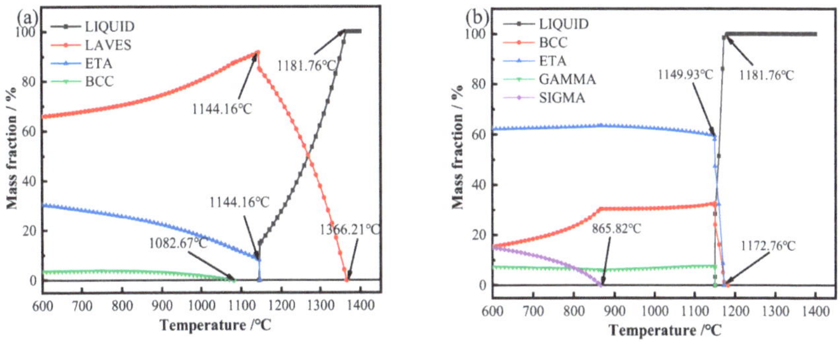

Figure 13. Equilibrium fraction of the composite at a deposition current of 40 A: (**a**) interdendritic and (**b**) dendrite axis.

Figure 14 shows the equilibrium fraction diagram of the composite material when the deposition current is 45 A. The calculation results show that the temperature at which all the dendrites are converted to the liquid phase is 1310.08 °C, and the temperature at which all the interdendrites are converted to the liquid phase is 1270 °C. It can be seen that during the cooling process, the γ-Ni phase solidifies first in the dendrite axis, and the temperature of the highest precipitation content of the γ-Ni phase is 1180.65 °C. The precipitation temperature of the Laves phase between dendrites is 1269.52 °C, which is similar to that of the γ-Ni phase. In the WAAM process, the high-temperature residence time is short, and the cooling rate is extremely fast, so during the cooling process, the γ-Ni phase and Laves phase precipitate almost simultaneously. With the decrease in temperature, the Ni_3Ti phase begins to precipitate on the dendrite axis. When the temperature further decreases, the γ-Ni content on the dendrite axis decreases, while the TiNi phase content increases. The main Ni_3Ti phase is precipitated between dendrites. When the temperature drops gradually, the content of the Laves phase decreases, and the content of the Ni_3Ti phase increases.

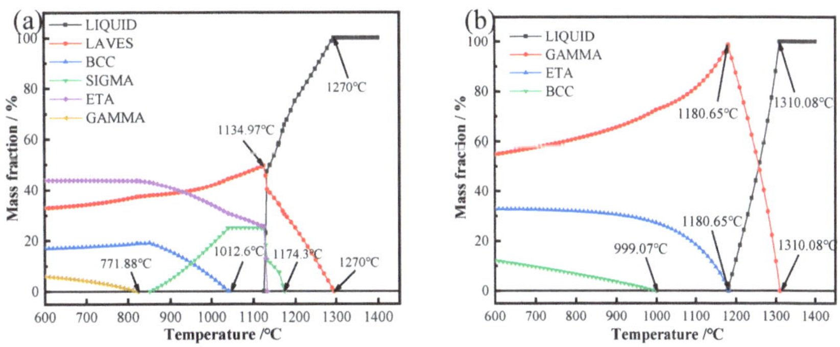

Figure 14. The equilibrium fraction of the composite at a deposition current of 45 A: (**a**) interdendritic and (**b**) dendrite axis.

Figure 15 presents the equilibrium fraction diagram of the composite at a deposition current of 50 A. The calculated results indicate that the precipitated phases include the Laves phase, Ni_3Ti phase, and TiNi phase, occurring both along the dendrite axes and in the interdendritic regions. The precipitation content of the Laves phase on the dendrite axis peaks at 1140 °C, subsequently decreasing as the temperature lowers. Conversely, the contents of the Ni_3Ti phase and TiNi phase gradually increase and approach a state of equilibrium.

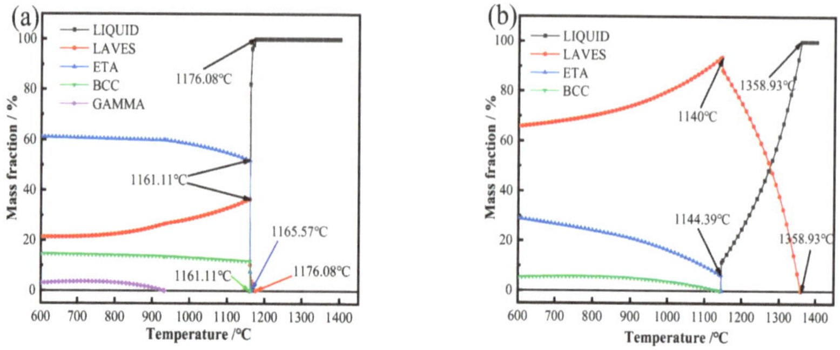

Figure 15. Equilibrium fraction of the composite with a deposition current of 50 A: (**a**) interdendritic and (**b**) dendrite axis.

The precipitation temperatures for the Laves, TiNi, and Ni_3Ti phases are approximately 1161.11 °C. As the temperature decreases, the content of the Laves phase tends to decrease and eventually reaches equilibrium, while the content of the Ni_3Ti phase slowly increases, also trending toward equilibrium. Throughout the solidification process from the liquid to solid state, the Laves phase on the dendrite axis preferentially precipitates, forming a liquid film between the dendrites, which helps maintain the dendritic structure.

4.2. Crack Analysis

The size of the welding heat input significantly influences the stress levels in the workpiece; generally, higher heat input leads to increased stress [40]. In this study, cracks occurred in the prepared composites at a deposition current of 50 A. As the deposition current increases, the high-temperature residence time is prolonged. SEM-EDS analysis of the composite material reveals the presence of MC carbides along the dendrite axis, indicating that these carbides precipitate during the solidification of liquid metal. The presence of MC carbides contributes to stress concentration in the workpiece, which can deteriorate its mechanical properties. Additionally, the presence of Laves phases also impacts the mechanical performance of the workpiece.

Microstructural analysis of the composites under different deposition currents reveals that at a deposition current of 50 A, there is a significant presence of Laves phases in both the interdendritic regions and along the dendrite axes. The presence of these Laves phases predisposes the workpiece to liquefaction cracking, adversely affecting its mechanical properties. Furthermore, SEM and EDS analyses of the fractured sections indicate that the failure mode is a cleave fracture, with a substantial presence of secondary phases detected at the fracture site. Solidification precipitation analysis suggests that these secondary phases primarily consist of the Laves phase, Ni_3Ti, and TiNi. XRD analysis also indicates the presence of Fe_2Ti and Cr_2Ti in the composite. It has been documented that the Fe_2Ti, Ni_3Ti, and Cr_2Ti phases are brittle intermetallic compounds [41,42]. The combined influence of the Laves phase and these secondary phases contributes to the crack formation in the composite.

4.3. Hardness Analysis

The microhardness of the composite increases with the increase in the deposition current. The secondary phase precipitation will also affect the microhardness of the composite. When the deposition current is small, the secondary phases such as Ti-Cr and Ti-Fe in the composite are less precipitated. When the deposition current increases, more secondary phases are generated, which increases the microhardness of the composite. In addition, the dendrites of the material are often filled with crystals with high solute content, resulting in segregation, which will affect the mechanical properties of the alloy, and the smaller the dendrite spacing, the more favorable the alloy properties [43]. The secondary dendrite spacing of composites produced under different deposition currents shows some variability. As illustrated in Figure 6, the secondary dendrite spacing initially increases with the deposition current and then decreases at higher currents. Concurrently, the microhardness of the composites also increases with the deposition current. At deposition currents of 35 A and 40 A, the secondary dendrite spacing increases, and the microhardness of the composites rises as well. This increase in microhardness at 40 A is attributed to a greater amount of secondary precipitates forming within the material, which enhances its properties. For deposition currents of 40 A and 45 A, the secondary dendrite spacing remains relatively constant, leading to similar microhardness values. However, when the deposition current reaches 50 A, the secondary dendrite spacing is reduced, and the precipitation of secondary phases is significantly higher, resulting in the highest microhardness observed.

5. Conclusions

1. The Ti6Al4V/Inconel 718 composite material was successfully fabricated using Ti6Al4V and Inconel 718 welding wires as raw materials, employing the double-wire Wire and Arc Additive Manufacturing (WAAM) deposition method at various deposition currents. However, it was observed that an increase in the deposition current correlated with the emergence of cracks within the composite material.
2. As the deposition current increases, the secondary dendrite spacing in the deposition samples first increases and then decreases. At a deposition current of 35 A, the secondary dendrite spacing measures 5.16 μm, while at 50 A, it reduces to 3.96 μm. Notably, the samples prepared at a deposition current of 35 A exhibit good morphology, with no visible cracks or defects.
3. At a deposition current of 35 A, the microstructure primarily consists of a nickel-based alloy, with Ti elements dispersed along the dendrite axis through diffusion. At deposition currents of 40 A and 45 A, the contents of titanium, nickel, chromium, and iron between the dendrite axis and the dendrites show minimal variation. However, at 50 A, TiNi, Cr_2Ti, and Fe_2Ti are predominantly deposited on the dendrite axis, while MC carbide is also found there. Additionally, Ni_3Ti is mainly deposited between the dendrites.
4. At a deposition current of 50 A, cracks appear in the prepared sample, exhibiting a cleavage fracture. The excessive heat input leads to stress concentration in the deposited layer, surpassing the yield strength of the alloy and resulting in a cracked cross-section.
5. The hardness data indicate that the microhardness of the sample increases with the current. Specifically, when the current rises to 50 A, the microhardness shows a 31.51% increase compared to the value at 35 A.

Author Contributions: Conceptualization, P.X. and X.Y.; methodology, P.X. and X.Y.; validation, P.X. and G.Z.; formal analysis, P.X.; investigation, P.X. and G.Z.; resources, X.Y.; data curation, P.X.; writing—original draft preparation, P.X.; writing—review and editing, P.X.; supervision, X.Y.; project administration, X.Y.; funding acquisition, X.Y. All authors have read and agreed to the published version of the manuscript.

Funding: This research is supported by a peak discipline grant.

Data Availability Statement: The original contributions presented in the study are included in the article, further inquiries can be directed to the corresponding authors.

Acknowledgments: This work was supported in part by Class III Peak Discipline of Shanghai Materials Science and Engineering (High-Energy Beam Intelligent Processing and Green Manufacturing).

Conflicts of Interest: The authors declare no conflict of interest.

References

1. Sakundarini, N.; Taha, Z.; Abdul-Rashid, S.H.; Ariffin, R.; Ghazila, R. Optimal multimaterial selection for lightweight design of automotive body assembly incorporating recyclability. *J. Mater. Des.* **2013**, *50*, 846–857. [CrossRef]
2. Wargnier, H.; Kromm, F.X.; Danis, M.; Brechet, Y. Proposal for a multi-material design procedure. *J. Mater.* **2014**, *56*, 44–49. [CrossRef]
3. Jeong, H.-I.; Kim, D.-H.; Lee, C.-M. Multi-material deposition of Inconel 718 and Ti-6Al-4V using the Ti-Nb-Cr-V-Ni high entropy alloy intermediate layer. *J. Mater. Res. Technol.* **2024**, *29*, 3217–3227. [CrossRef]
4. Hou, X.; Ye, X.; Qian, X.; Zhang, X.; Zhang, P.; Lu, Q.; Yu, Z.; Shen, C.; Wang, L.; Hua, X. Heat Accumulation, Microstructure Evolution, and Stress Distribution of Ti-Al Alloy Manufactured by Twin-Wire Plasma Arc Additive. *J. Adv. Eng. Mater.* **2022**, *24*, 2101151. [CrossRef]
5. Hadibeyk, S.; Beidokhti, B.; Sajjadi, S.A. Effect of bonding time and homogenization heat treatment on the microstructure and mechanical properties of the transient liquid phase bonded dissimilar GTD-111/FSX-414 TLP superalloys. *J. Alloy. Comp.* **2018**, *731*, 929–935. [CrossRef]
6. Zhang, Y.; Luo, Z.; Li, Y.; Liu, Z.; Huang, Z. Microstructure characterization and tensile properties of Mg/Al dissimilar joints manufactured by thermocompensated resistance spot welding with Zn interlayer. *J. Mater.* **2015**, *75*, 166–173. [CrossRef]
7. Aceves, S.M.; Espinosa-loza, F.; Elmer, J.W.; Huber, R. ScienceDirect Comparison of Cu, Ti and Ta interlayer explosively fabricated aluminum to stainless steel transition joints for cryogenic pressurized hydrogen storage. *J. Int. J. Adv. Manuf. Technol.* **2014**, *40*, 1490–1503. [CrossRef]
8. Oliveira, J.P.; Panton, B.; Zeng, Z.; Andrei, C.M.; Zhou, Y.; Miranda, R.M.; Fernandes, F.M.B. Laser joining of NiTi to Ti6Al4V using a Niobium interlayer. *J. Acta Mater.* **2016**, *105*, 9–15. [CrossRef]
9. Mishra, A.; Paul, A.R.; Sharma, R.; Mukherjee, M.; Singh, R.K. Interfacial characteristics of Ti6Al4V-IN718 dissimilar structure developed by wire-arc additive manufacturing using Monel-400 as an interlayer. *J. Mater. Today Proc.* **2023**, *80*, 241–247. [CrossRef]
10. Song, C.; Ma, G.; He, A.; Li, Y.; Niu, F.; Wu, D. Multi-phases transformation mechanism of Ti6Al4V/Inconel 718 composite by laser additive manufacturing. *J. Mater. Charact.* **2021**, *179*, 111363. [CrossRef]
11. Chatterjee, S.; Abinandanan, T.A.; Chattopadhyay, K. Phase formation in Ti/Nidissimilar welds. *J. Mater. Sci. Eng. A* **2008**, *490*, 7–15. [CrossRef]
12. Chatterjee, S.; Abinandanan, T.A.; Reddy, G.M.; Chattopadhyay, K. Microstructure formation in dissimilar metal welds: Electron beam welding of Ti/Ni. *J. Metall. Mater. Trans. A* **2016**, *47*, 769–776. [CrossRef]
13. Martinsen, K.; Hu, S.J.; Carlson, B.E. Joining of dissimilar materials. *J. CIRP Ann.-Manuf. Technol.* **2015**, *64*, 679–699. [CrossRef]
14. Sun, Z.; Ion, J.C. Laser welding of dissimilar metal combinations. *J. Mater. Sci.* **1995**, *30*, 4205–4214. [CrossRef]
15. Wang, Y.; Ye, X.; Shi, M.; Pan, N.; Xia, P. Study of Phase Evolution Behavior of Ti6Al4V/Inconel 718 by Pulsed Laser Melting Deposition. *J. Mater.* **2023**, *16*, 2437. [CrossRef]
16. Chen, H.C.; Pinkerton, A.J.; Li, L. Fibre laser welding of dissimilar alloys of Ti-6Al-4V and Inconel 718 for aerospace applications. *J. Int. J. Adv. Manuf. Technol.* **2011**, *52*, 977–987. [CrossRef]
17. Peng, Y.; Fu, Z.Y.; Wang, W.M.; Zhang, J.Y.; Wang, Y.C.; Wang, H.; Zhang, Q.J. Welding of Ti-6Al-4V and TiB2-Ni cermet using pulsed current heating. *J. Sci. Technol. Weld. Join.* **2008**, *13*, 456–461. [CrossRef]
18. Sun, Z.; Ji, X.; Zhang, W.; Chang, L.; Xie, G.; Chang, H.; Zhou, L. Microstructure evolution and high temperature resistance of Ti6Al4V/Inconel625 gradient coating fabricated by laser melting deposition. *J. Mater. Des.* **2020**, *191*, 108644. [CrossRef]
19. Meng, W.; Xiaohui, Y.; Zhang, W.; Junfei, F.; Lijie, G.; Qunshuang, M.; Bing, C. Additive manufacturing of a functionally graded material from Inconel625 to Ti6Al4V by laser synchronous preheating. *J. Mater. Process. Technol.* **2020**, *275*, 116368. [CrossRef]
20. Bonny, O.; Amit, B. Additive manufacturing of Inconel 718-Ti6Al4V bimetallic structures. *J. Addit. Manuf.* **2018**, *22*, 844–851. [CrossRef]
21. Huang, J.; Liu, G.; Yu, X.; Guan, Z.; Yu, S.; Fan, D. Characterization of nickel-titanium alloy graded materials using double wire alternating current cross arc additive manufacturing. *J. Alloys Compd.* **2022**, *910*, 164912. [CrossRef]
22. Lu, L.; Tian, Y.; Cai, Y.; Xin, Y.; Chen, X.; Zhang, G.; Han, J. Microstructure and mechanical properties of a functionally graded material from TA1 to Inconel 625 fabricated by dual wire+arc additive manufacturing. *J. Mater. Lett.* **2021**, *298*, 130010. [CrossRef]
23. Wang, J.; Pan, Z.; Wei, L.; He, S.; Cuiuri, D.; Li, H. Introduction of ternary alloying element in wire arc additive manufacturing of titanium aluminide intermetallic. *J. Addit. Manuf.* **2019**, *27*, 236–245. [CrossRef]
24. Han, J.; Chen, X.; Zhang, G.; Liu, B.; Cai, Y.; Chen, M.; Jiang, H.; Tian, Y. Ni50.8Ti49.2 alloy prepared by double-wire + arc additive manufacturing with a substrate heating temperature of 600 C. *J. Manuf. Process.* **2023**, *104*, 76–86. [CrossRef]
25. Le, V.T.; Bui, M.C.; Nguyen, T.D.; Nguyen, V.A. Van Canh Nguyen. On the connection of the heat input to the forming quality in wire-and-arc additive manufacturing of stainless steels. *J. Vacuum.* **2023**, *209*, 111807. [CrossRef]
26. Le, V.T.; Mai, D.S.; Bui, M.C.; Wasmer, K.; Nguyen, V.A.; Dinh, D.M.; Nguyen, V.C.; Vu, D. Influences of the process parameter and thermal cycles on the quality of 308L stainless steel walls produced by additive manufacturing utilizing an arc welding source. *J. Weld World* **2022**, *66*, 1565–1580. [CrossRef]
27. Zhang, G.; Chen, X.; Cai, Y.; Lu, L.; Xin, Y.; Jia, D.; Tian, Y.; Han, J. Microstructure and mechanical properties of Ti-3Al-2V alloy by dual wire + arc additive manufacturing. *J. Mater. Lett.* **2021**, *229*, 130190. [CrossRef]
28. Danek, M.; Fernandes, F.; Cavaleiro, A.; Polcar, T. Influence of Cr additions on the structure and oxidation resistance of multilayered TiAlCrN films. *J. Surf. Coat. Technol.* **2017**, *313*, 158–167. [CrossRef]

29. Hu, B.; Du, Y.; Schuster, J.C.; Sun, W.; Liu, S.; Tang, C. Thermodynamic modeling of the Cr-Ni-Ti system using a four-sublattice model for ordered/disordered bcc phases. *J. Thermochim. Acta* **2014**, *578*, 35–42. [CrossRef]
30. Górka, J.; Jamrozik, W.; Kiel-Jamrozik, M. The effect of TIG welding on the structure and hardness of butt joints made of Inconel 718. *J. Heliyon.* **2023**, *9*, e13175. [CrossRef]
31. Tian, Y.; Chen, X.; Cai, Y.; Luo, Z.; Chen, M.; Zhang, X.; Li, J.; Han, J. Microstructure and properties of a Ni-Ti-Cr-Mo-Nb alloy fabricated in situ by dual-wire arc additive manufacturing. *J. Mater. Sci. Eng. A* **2022**, *853*, 143740. [CrossRef]
32. Wu, Y.; Wu, X.; Qin, S.; Yang, K. Compressibility and phase transition of intermetallic compound Fe2Ti. *J. Alloys Compd.* **2013**, *558*, 160–163. [CrossRef]
33. Uchil, J.; Kumara, K.G.; Mahesh, K.K. Effects of heat treatment temperature and thermal cycling on phase transformations in Ni-Ti-Cr alloy. *J. Alloys Compd.* **2001**, *325*, 210–214. [CrossRef]
34. Saunders, N.; Miodownik, A.P. *Calphad-Calculation of Phase Diagrams: A Comprehensive Guide*; Pergamon materials series; Cahn, R.W., Ed.; Elsevier: Amsterdam, The Netherlands, 1998; Volume 1, pp. 1–479.
35. Shercliff, H.R.; Ashby, M.F. A process model for age hardening of aluminum alloys-I. the model. *J. Acta Metall. Et Mater.* **1990**, *38*, 1789–1802. [CrossRef]
36. Yang, Z.; Yan, C.; Tang, T.; Li, P.; Li, Z. Failure analysis and heat treatment process optimization of NOS525 rotary flat binaural hot forging die. *J. Eng. Fail. Anal.* **2024**, *163*, 108521. [CrossRef]
37. Swartzendruber, L.J.; Itkin, V.P.; Alcock, C.B. The Fe-Ni (iron-nickel) system. *J. Phase Equilibria* **1991**, *12*, 288–312. [CrossRef]
38. Ojo, O.A. Intergranular liquation cracking in heat affected zone of a welded nickel based superalloy in as cast condition. *J. Mater. Sci. Technol.* **2007**, *23*, 1149–1155. [CrossRef]
39. Chintababu, U.; Chary, V.R.; Gupta, S.P. Liquid Film Migration in a Cu-5 At.% Sb Alloy. *J. Can. Metall. Q.* **2007**, *46*, 175–187. [CrossRef]
40. Kumar, S.; Shahi, A.S. Effect of heat input on the microstructure and mechanical properties of gas tungsten arc welded AISI 304 stainless steel joints. *J. Mater. Design.* **2011**, *32*, 3617–3623. [CrossRef]
41. Pelloth, J.; Brand, R.A.; Keune, W. Local magnetic properties of the Fe2Ti Laves phase. *J. Magn. Magn. Mater.* **1995**, *140–144*, 59–60. [CrossRef]
42. Su, L.; Jia, L.; Feng, Y.; Zhang, H.; Yuan, S.; Zhang, H. Microstructure and room-temperature fracture toughness of directionally solidified Nb-Si-Ti-Cr-Al-Hf alloy. *J. Mater. Sci. Eng. A Struct. Mater. Prop. Microstruct. Process.* **2013**, *560*, 672–677. [CrossRef]
43. Liu, M.Y.; Shi, B.; Wang, C.; Ji, S.K.; Cai, X.; Song, H.W. Normal Hall–Petch behavior of mild steel with submicron grains. *J. Mater. Lett.* **2003**, *57*, 2798–2802. [CrossRef]

materials

Article

Study on Surface Quality Analysis of an Uncoated Boron Steel and Its Oxide Layer Suppression Method for Hot Stamping

Jiho Lee, Junghan Song and Gihyun Bae *

Flexible Manufacturing R&D Department, Korea Institute of Industrial Technology, Incheon 21999, Republic of Korea; jksg0319@kitech.re.kr (J.L.); jhsong@kitech.re.kr (J.S.)
* Correspondence: baegh@kitech.re.kr; Tel.: +82-32-850-0318

Abstract: This study investigates the effects of hot stamping on boron steel surface properties, comparing uncoated steel to Al–Si-coated steel, with a focus on developing atmosphere-controlled hot stamping technology. Experiments using a hat-shaped specimen revealed that uncoated steel formed a thick oxide layer due to exposure to atmospheric oxygen at high temperatures, negatively impacting surface quality and weldability. In contrast, the Al–Si-coated steel showed no oxide formation. Although uncoated steel exhibited higher average Vickers hardness, the detrimental effects of the oxide layer on weld quality necessitate advancements in process technology. A lab-scale hot stamping simulator was developed to control atmospheric oxygen levels, utilizing a donut-shaped induction heating coil to heat the material above 1000 °C, followed by rapid cooling in a forming die. Results demonstrated that maintaining oxygen concentrations below 6% significantly reduced oxide layer thickness, with near-vacuum conditions eliminating oxide formation altogether. These findings emphasize the critical role of oxygen control in enhancing the surface quality and weldability of uncoated boron steel for ultra-high-strength automotive applications, potentially reducing manufacturing costs while ensuring part performance.

Keywords: hot stamping; boron steel sheet; oxide layer; atmosphere control; induction heating module

Citation: Lee, J.; Song, J.; Bae, G. Study on Surface Quality Analysis of an Uncoated Boron Steel and Its Oxide Layer Suppression Method for Hot Stamping. *Materials* **2024**, *17*, 5563. https://doi.org/10.3390/ma17225563

Academic Editors: Dimitrios Tzetzis, Konstantinos Tsongas and Nikolaos Kladovasilakis

Received: 18 October 2024
Revised: 11 November 2024
Accepted: 12 November 2024
Published: 14 November 2024

1. Introduction

Due to recent environmental concerns and strengthened fuel efficiency regulations, the demand for eco-friendly vehicles is increasing. Many global automotive companies are producing electric vehicles as a means to satisfy various environmental regulations by replacing traditional internal combustion engines. In order to achieve maximum driving range with a single battery charge, the weight of the vehicle has become crucial for environmentally friendly vehicles. To reduce vehicle weight, many companies are increasing the application of hot stamping components to ensure collision performance and passenger safety. As evidence of this, after the initial application of hot stamping components made of boron steel in a Saab vehicle in 1984 [1], the production ratio of hot stamping increased from 3 million to 8 million in 1987 and further surged to 170 million in 2007 [2]. In 2023, the hot stamping market is expected to expand to approximately 1.28 billion components. Accordingly, research related to hot stamping production technology is actively being conducted.

The hot stamping method is suitable for producing high-strength products required for passenger safety by forming the material with high elongation and low strength under a high temperature of over 900 °C and then quenching it to the martensitic steel in order to achieve very high strength over 1.5 GPa. The quenching process involves rapid cooling at speeds exceeding approximately 27 K/s from the martensite transformation temperature, ranging from 425 °C (martensite start point) to 280 °C (martensite finish point) [3]. Materials commonly used for hot stamping include alloys with boron additions such as 22MnB5, 27MnB5, and 37MnB5. Boron-added alloys are preferred in hot stamping, as boron steel is known to be the only type of steel capable of generating a martensitic microstructure after hot stamping [4]. Boron steel sheets typically have tensile strength levels of around 600 MPa before heat treatment. Through the martensitic transformation of the material after hot stamping, it is known to achieve tensile strengths of approximately 1.5 GPa [4].

Numerous researchers have investigated component manufacturing technologies utilizing uncoated hot stamping steel sheets. From a materials technology perspective, various experimental analyses have been conducted to ensure and enhance the mechanical properties of materials, with the cooling rate selected as a key variable [5–7]. Concurrently, from a process technology perspective, researchers have focused on optimizing the hot stamping process to improve forming quality and productivity [8–12]. Furthermore, a variety of studies have been conducted to analyze heat transfer and material deformation mechanisms in the hot stamping process, aiming to enhance process efficiency [13–16]. These collective studies have continuously advanced component manufacturing technologies that utilize uncoated hot stamping steel sheets.

Hot-stamped boron steels without Al–Si coating layer generate an oxide scale on the surface when exposed to air under austenitizing conditions. To control this oxide layer, methods such as using boron steel sheets coated with Al–Si for oxidation prevention [17] or applying oil for anti-oxidation on the surface [18] can be employed. Physical methods involving shot blasting can also remove the oxide layer [18,19]. While these methods offer the advantage of controlling the oxide layer during hot stamping, the presence of a coating layer on the surface for oxidation control can lead to the melting of the coating layer during welding. Additionally, when using oil for oxidation prevention, research data indicates a decrease in interfacial heat transfer coefficients (IHTC), affecting the microstructure and mechanical properties [20]. The Al–Si coating layer, when subjected to rapid heating rates, can experience premature heating, causing the coating layer to slide and leading to issues such as coating layer displacement. Moreover, the production cost of hot-stamped parts should be increased due to material and process patents related to Al–Si-coated boron steel sheets.

To overcome these drawbacks, it is important to suggest an alternative hot stamping method by using an uncoated boron steel sheet. Taylor et al. [21] illustrated a schematic of the hot stamping process for uncoated steel sheets, demonstrating that this process generally includes an additional shot blasting step to remove the oxidation scale. This additional step requires extra production equipment and extends the processing time, resulting in increased manufacturing costs for hot-stamped parts using uncoated steel sheets. The comparatively higher manufacturing cost, as opposed to coated steel sheets, presents a significant obstacle to the use of non-coated steel sheets. Consequently, even with the additional costs associated with paying for material and process patents held by Arcelor Mittal, most automotive body component manufacturers opt to use Al–Si-coated steel sheets. Chen et al. [22] emphasized the cost-saving advantages of uncoated hot stamping processes in automotive part manufacturing. Additionally, Kim et al. [23] demonstrated through an economic efficiency analysis that using uncoated steel sheets in hot stamping processes can reduce production costs and improve efficiency. Therefore, to reduce the manufacturing cost of hot-stamped components, automotive companies and body component manufacturers are seeking the development of press-forming technology that can suppress the formation of oxide layers on non-coated steel sheets. It means that preventing or controlling oxide layer growth should be a core technology for using uncoated boron steel sheets.

This paper deals with the quantitative investigation of the surface quality of a hot-stamped part and its improvement method of the oxide layer for uncoated boron steel sheet. Initially, the heat treatment results with hat-type specimen were compared according to the presence of the Al–Si coating layer in order to observe the effect of the coating layer on the surface quality and mechanical properties. To investigate methods for suppressing the oxide layer on uncoated steel sheets, a lab-scale hot stamping simulator was developed to replicate the heating and forming processes under controlled atmospheric conditions.

Subsequently, the oxide layers were quantitatively analyzed according to the oxygen ratio. The research results can be utilized as hot stamping process control technology to ensure surface quality and weldability in the manufacturing process of ultra-high-strength body components using uncoated steel sheets.

2. Hot Stamping Test Method

2.1. Equipment Setup

Al–Si-coated boron steel is known to prevent the formation of an oxide layer by blocking the oxygen contact during the hot stamping process [17]. For using uncoated boron steel sheets, the surface quality and property should be investigated to observe the oxide layer formation and to ensure the mechanical properties of hot-stamped parts.

An experimental setup, as shown in Figure 1, is constructed in order to reproduce the hot stamping process. An atmosphere furnace has the capability of heating up to approximately 1000 °C, and a mechanical press can make hat-type specimens by using a hot stamping die. A water-cooling chiller is connected to the stamping die for rapid cooling, and a thermometer is used to monitor the temperature at the stamping die. The 300 mm × 300 mm sample was heated up to 950 °C and stabilized for 10 min. After that, they rapidly cooled down to 10 °C or below within 5 s for the martensite phase transformation in the die. 22MnB5 1.0 t steel sheets, produced by Arcelor Mittal, with and without Al–Si coating layer, were used for experiments. The chemical composition and the mechanical strength of 22MnB5 base metal are presented in Table 1.

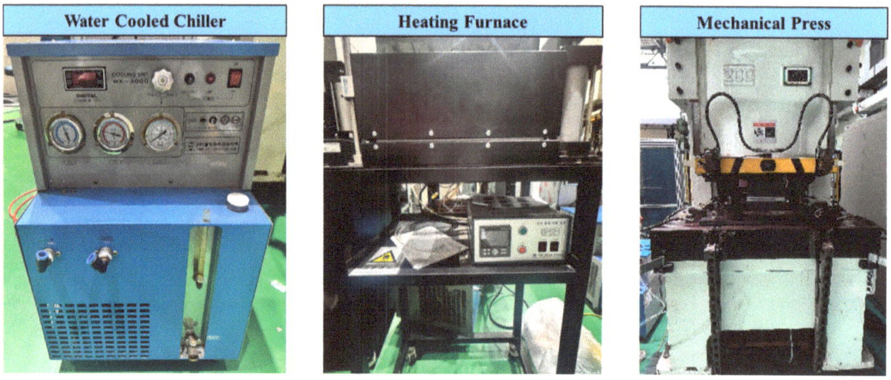

Figure 1. Main test devices for the hot stamping test.

Table 1. Chemical compositions and mechanical strength of 22MnB5 base metal.

Chemical Compositions [wt%]							Mechanical Strength [MPa]	
C	Mn	B	Si	P	S	Cr	Yield Stress	Tensile Stress
0.24	1.10	0.003	0.26	0.014	0.001	0.14	1254	1545

2.2. Test Procedure

To observe the surface property changes after hot stamping for both uncoated and Al–Si-coated materials, specimens were subjected to the same heat treatment conditions. The specimens were heated by the controllable atmosphere furnace up to 950 °C and stabilized for 10 min. Subsequently, they were quenched and formed simultaneously using a 200-ton servo press with a water-cooled die that was pre-cooled to below 10 °C. The time taken for the material to reach the press was approximately 5 s. The detailed test procedure is presented in Figure 2.

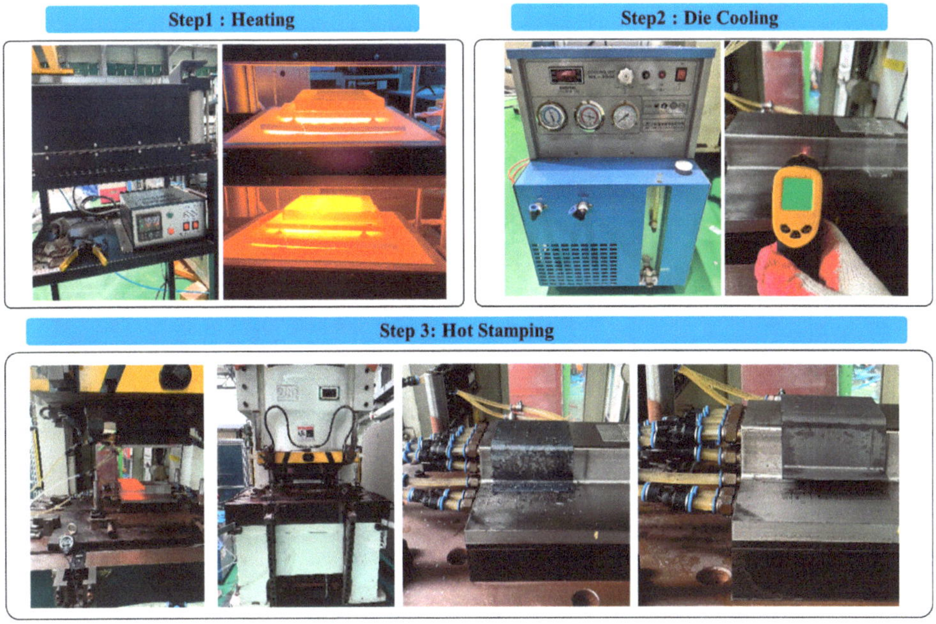

Figure 2. Hot stamping test process of a hat-type specimen.

3. Hot Stamping Test Results

3.1. Surface Investigation

Hot stamping experiments were performed with hat-type specimens in order to observe the growth of the oxide layer and to investigate the surface property of the coated and uncoated boron steel sheet. Figure 3 shows test specimens according to the coating layer. The specimen with Al–Si coating layer shows only the heat-affected color change due to the rapid temperature change during the heating and quenching process, and no oxide layer was observed as shown in Figure 3a. This aligns with the known characteristic of Al–Si coating preventing the formation of oxide layers by blocking contact with the atmosphere after heating [17]. As shown in Figure 4, the heated material generates an oxide layer due to contact with oxygen in the air during the transport process, and after rapid cooling during the component forming process, a thick oxide layer exists as shown in Figure 3b.

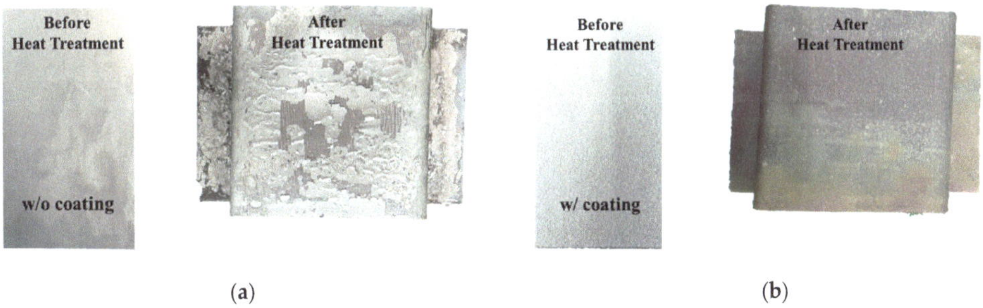

Figure 3. Hat-type specimens before and after the hot stamping test: (**a**) uncoated material; (**b**) coated material.

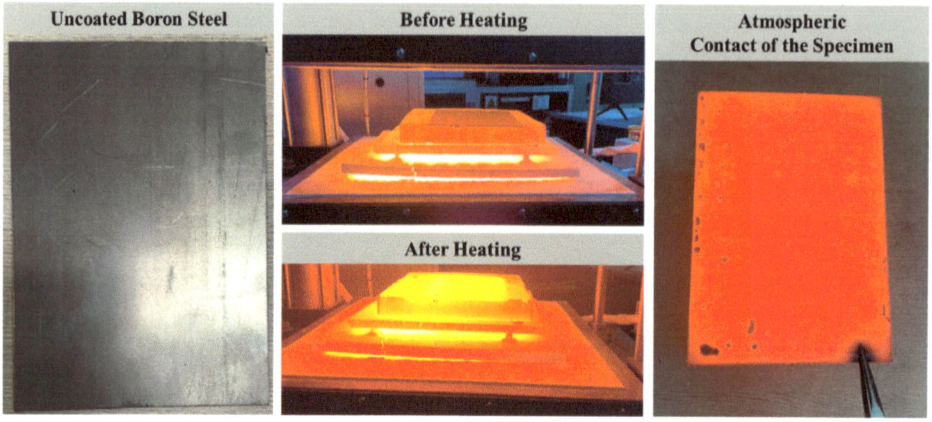

Figure 4. Oxide layer formation of the uncoated boron steel due to oxygen contact.

3.2. Oxide Layer

For a more precise analysis of surface layer changes, the SEM (scanning electron microscope) equipment was employed to observe coated and uncoated boron steel sheets before and after heating. The equipment used for observations is the QUANTA 200F (FEI Company, Hillsboro, OR, USA). First of all, point EDS measurements were carried out to investigate the composition of the surface scale of the uncoated boron steel sheet. Figure 5 shows the result of point EDS measurements of the surface scale. The composition of the surface scale was Fe 80.74 wt% and O 19.26 wt% components, and it reveals that the surface scale is the oxide layer of Fe and O.

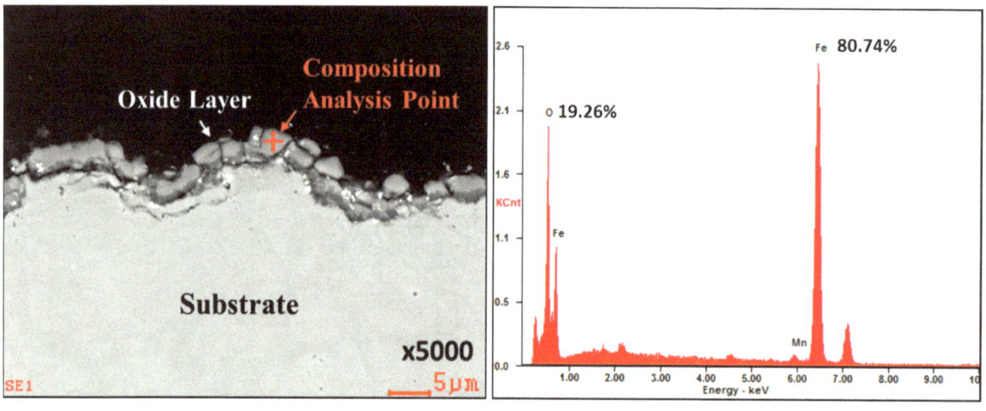

Figure 5. Point EDS composition analysis of uncoated boron steel sheets after heat treatment.

The Al–Si-coated specimen was analyzed using EDS mapping in order to investigate the change in the composition at the coating layer before and after heat treatment. As shown in Figure 6, Al–Si coating layer is transformed into the mixed intermetallic layer of Al–Si and Fe after the heat treatment. As shown in Figure 6, fine cracks and internal voids are observed in the rough intermetallic layer, which are generated during the rapid quenching process. Such compounds are known to have low fracture toughness, promoting the formation and growth of fine cracks [24,25]. As a result of the diffusion and alloying of components, the coating thickness is increased from 19.79 μm to 38.14 μm after heat treatment.

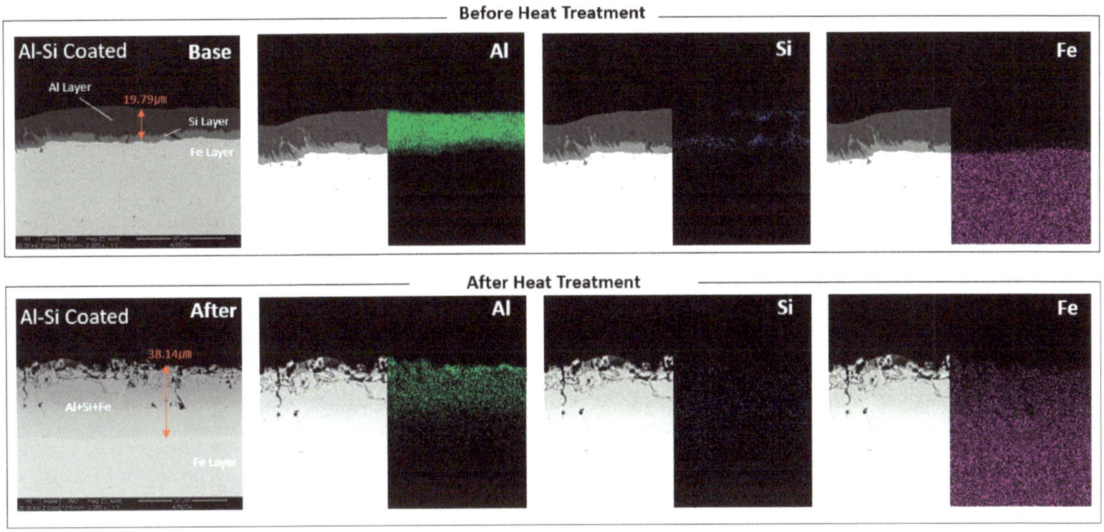

Figure 6. EDS mapping analysis of Al–Si-coated boron steel sheets before and after heat treatment.

The oxide layer thickness was quantitatively measured by optical microscope with a hat-type specimen of uncoated boron steel sheet in order to investigate the oxide layer residual due to the different contact mechanisms between die and sheet during the hot stamping. Figure 7 shows the captured location of samples from the hat-type uncoated specimen. As shown in Figure 8, the oxide layer thickness was 23.46 μm, 1.81 μm, and 7.55 μm at the top, wall, and flange locations, respectively. At the top location, the thickest oxide layer was observed due to no contact between the die and the sheet. On the contrary, most of the oxide layer was removed in the wall location because severe contact occurs when the sheet slides against the die during the forming. At the flange location, the thickness is decreased due to the high compression of the die at the bottom dead center of the press slide. Most spot welds are made at the top and flange areas in the auto-body assembly process. It means that the thick oxide layer can deteriorate the welding quality by using an uncoated boron steel sheet in the hot stamping. Therefore, a novel stamping method is required to reduce or control the oxide layer formation in order to utilize the uncoated boron steel sheet for ultra-high-strength automotive parts.

Figure 7. Sample-capturing locations of the hat-type specimen.

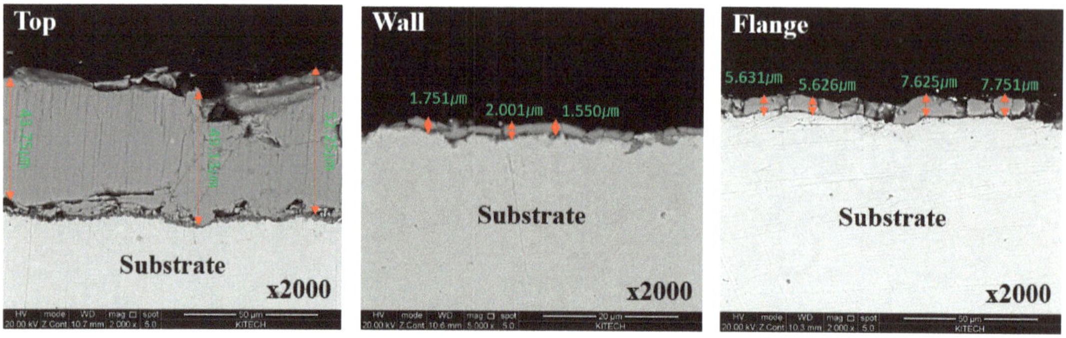

Figure 8. SEM images at the top, wall, and flange locations to measure the oxide layer thickness of uncoated boron steel.

The optical microscope images of Al–Si-coated specimen are presented in Figure 9. Because of the Al–Si coating layer, increased coating thickness according to the diffused alloying components is observed instead of the oxide layer. Additionally, microcracks and pores are present within the coating due to the different coefficients of thermal expansion [25–27]. No significant difference was observed at the top, wall, and flange locations due to the contact pressure during the forming process.

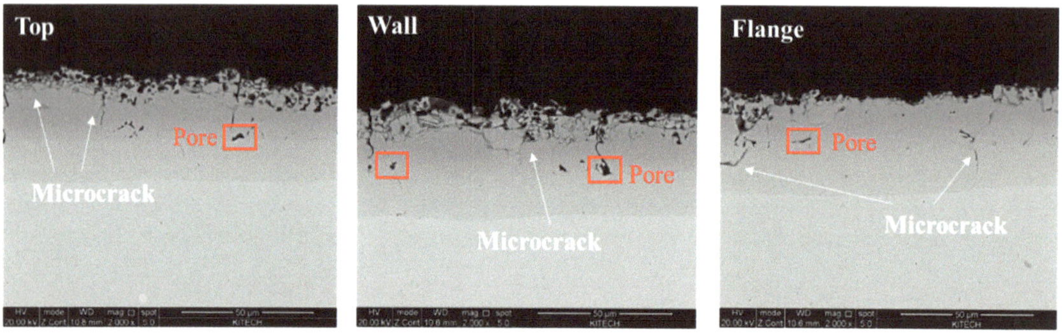

Figure 9. SEM images of Al–Si-coated hat-type specimen at the top, wall, and flange locations.

3.3. Hardness

A Vickers hardness test was conducted to examine the effect of the absence of a coating layer on material strength during the forming process of uncoated steel sheets. Hardness is generally known to correlate with material strength. Kim et al. [28] reported that the Vickers hardness of a boron steel sheet is approximately 450 Hv after the hot stamping thermal cycle. Based on phase transformation tests, Li et al. [29] demonstrated that the Vickers hardness of boron steel is 459 Hv at a cooling rate of 30 °C/s and that the hardness increases proportionally with the rise in cooling rate. In this paper, Vickers hardness was measured using a SHIMAZU HMV-2R (Shimadzu Corporation, Kyoto, Japan) device under the condition of applying a load of 980.7 mN for 10 s. Figure 10 shows the Vickers hardness measurement results for both coated and uncoated steel sheets. To examine the effect of cooling pattern changes due to differences in contact conditions between the material and the die, hardness was measured in the top, wall, and flange areas. The hardness was measured seven times at 1 mm intervals. The overall average hardness of the coated and uncoated steel sheets was measured at 481 Hv and 513 Hv, respectively, with higher values observed for the uncoated steel sheets. Therefore, it was confirmed that uncoated materials are more advantageous for strength securement in hot stamping, which requires rapid cooling. The hardness differences at various measurement locations were not considered significant when taking into account the scatter in the data, suggesting that the impact of the contact conditions between the die and the material on hardness is minimal. Despite the differences due to the presence or absence of a coating layer and contact conditions, the hat-type specimen can achieve the desired tensile strength in all areas.

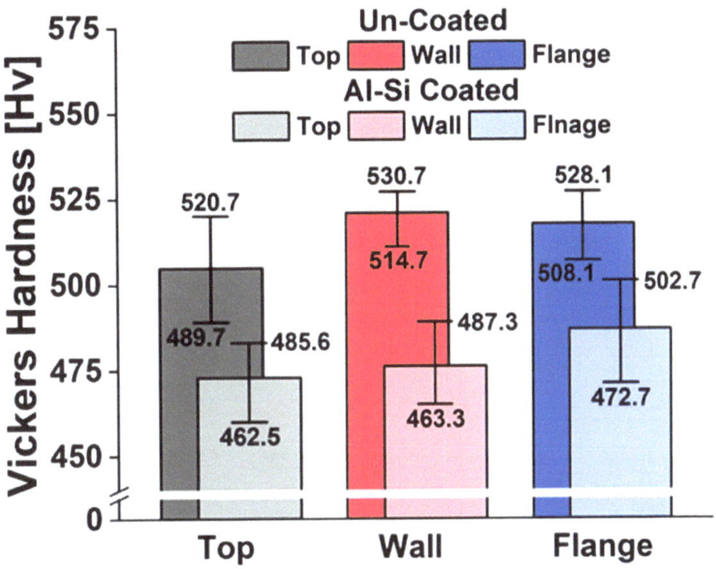

Figure 10. Vickers hardness of uncoated and coated boron steel sheets.

3.4. Summary

For the boron steel with Al–Si coating, as expected, no oxide layer was observed on the surface during the heat treatment and hot stamping processes. However, microcracks and pores were identified within the coating layer due to the different thermal expansion coefficients. Additionally, the elements clearly distinguished in the coating layer before heat treatment diffused during the heat treatment, forming a Fe_3(AlSiFe) alloy layer inside the coating layer, which could potentially compromise the weldability. The coating layer thickness also increased approximately two fold. On the other hand, the uncoated boron steel exhibited irregular and uneven oxide layers on the surface when exposed to the atmosphere during the heat treatment process. Hot stamping simulation tests revealed that the oxide layer thickness could be preserved, peeled off, or compressed according to the die shape. The oxide layer thickness was observed to be over 20 μm, which could cause weldability issues in the assembly process of automotive parts. The use of uncoated steel sheets is expected to be advantageous for securing material strength.

During the hot stamping process using uncoated 22MnB5 material, the oxidation reaction was most pronounced at the point of material transfer. This observation indicated that the material, upon being heated and transferred, came into contact with atmospheric oxygen, leading to the initiation and rapid growth of the oxide layer. Therefore, to ensure the surface quality of uncoated 22MnB5 material during hot stamping, controlling the contact between the heated material and oxygen during transfer is crucial for maintaining surface quality. Therefore, further research on new forming technologies is needed to suppress the formation of oxide layers in order to ensure the surface quality and weldability of uncoated steel sheets.

4. Oxide Layer Improvement

4.1. Development of a New Hot Stamping Simulator

The issue of oxide layer formation during the hot stamping process of uncoated steel sheets occurs because the material rapidly reacts with oxygen in the atmosphere at high temperatures, leading to the formation of an oxide layer. In particular, controlling the oxygen concentration in the atmosphere is crucial to suppressing the formation of the oxide layer. Developing a new lab-scale simulator that can replicate the hot stamping process is a very important approach to achieve this.

A lab-scale simulator allows for a quantitative analysis of the oxide layer thickness under various conditions, such as oxygen concentration and temperature, providing critical data for process optimization. Figure 11 shows the schematic diagram of the hot stamping simulator. The experimental equipment consists of a heating zone and a forming zone, including a transfer system for continuous forming after heating. To control the oxygen content in the atmosphere during the experiment, the entire equipment is sealed within a vacuum chamber. A nitrogen gas injection system was introduced to enhance the efficiency of atmospheric control. With this configuration, the vacuum chamber and nitrogen gas injection system allow the setting of controlled atmospheric conditions, making it possible to simulate the hot stamping process of uncoated steel sheets through the heating-transfer-forming sequence.

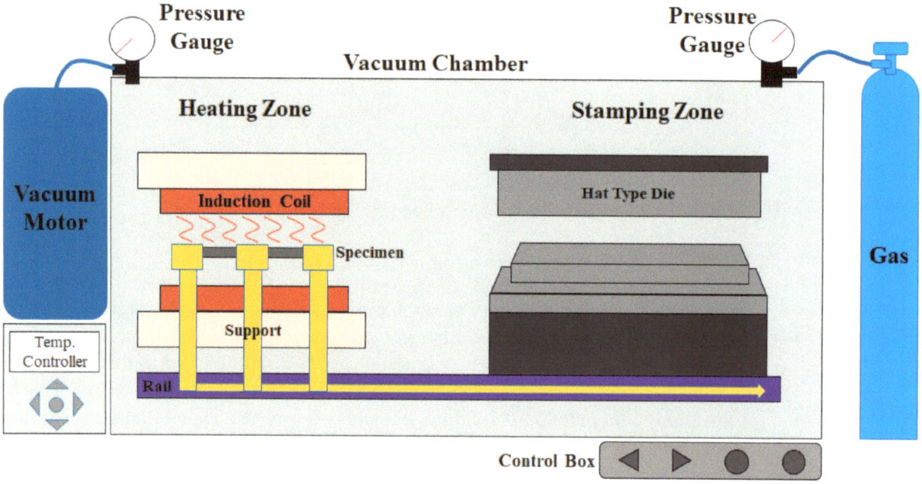

Figure 11. Schematic diagram of an atmosphere-controllable hot stamping simulator.

Figure 12 shows the manufactured experimental equipment. In the heating zone, a 300 mm × 300 mm specimen is placed on a transfer jig and heated using a pancake-shaped induction coil installed above it. Once the specimen reaches the target temperature, it is transferred to the stamping zone via a transfer device operated by a mechanical motor, where it is then formed by an upper die driven by a hydraulic cylinder. During this process, both the upper and lower dies are cooled with water, creating rapid cooling conditions that promote martensitic structure formation. For oxidation layer control experiments, an atmosphere control system was implemented as previously described, using a vacuum system to evacuate air from the chamber while injecting nitrogen gas to quickly reach the desired oxygen level. Using this newly developed hot stamping simulator, forming tests on non-coated hot stamping steel sheets were conducted at different oxygen levels to study the potential for oxidation layer suppression.

Figure 12. Lab-scale hot stamping simulator.

4.2. Establishment of the Experimental Condition

The simulator equipment uses a pancake-shaped induction coil to heat material of 300 mm × 300 mm size to temperatures above 950 °C. Due to the shape of the coil and the thermal deformation of the specimen, temperature variations occur at different positions. Therefore, a heating test was conducted to select the optimal output of the heating module for achieving the target temperature and to determine the analysis area for surface layer examination based on temperature variations. To measure the temperature distribution of the specimen, a FLIR A700 (FLIR Systems, Wilsonville, OR, USA) infrared camera capable of measuring temperatures up to 2000 °C was used. Figure 13 shows the heating performance test of the material using the FLIR camera. First, the material was heated under a 12 kW condition, and Figure 14 illustrates the distribution of the oxide layer formed after the heating test. Through this preliminary heating test, it was confirmed that the material was heated in line with the shape of the coil and that the oxide layer formed accordingly. The FLIR camera was used to select 1/4 of the dome-shaped specimen for measurement, considering the measurement range. Three thermocouples were attached to the wall section and flange section to measure the reference temperature for calibration.

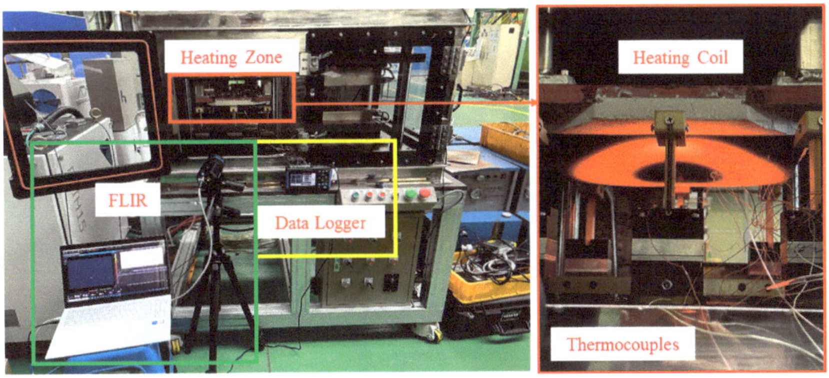

Figure 13. Equipment configuration for the performance evaluation of the heating module.

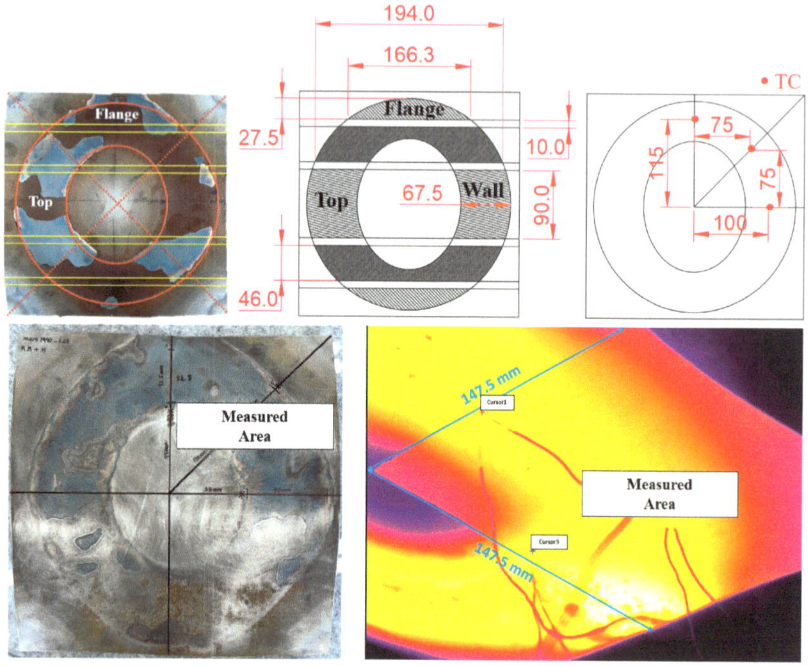

Figure 14. Locations of thermocouple attachment considering the heated area.

Afterward, heating tests were conducted according to the output of the heating module, and the temperature distribution was observed using the FLIR camera. Figure 15 shows the temperature distribution map corresponding to the heating module output. Since the target temperature for the hot stamping process is above 950 °C, this area was marked in yellow on the temperature distribution map. Through the heating test, it was confirmed that an output of 20 kW is required to maintain the target temperature across a wide area of the hat-type specimen. Therefore, the optimal output for forming tests under controlled oxygen conditions was set to 20 kW.

To verify the atmosphere control performance of the vacuum chamber, the changes in oxygen content according to nitrogen gas pressure were observed. The oxygen content was controlled by simultaneously removing air from the chamber using a vacuum pump and injecting nitrogen gas. To check the oxygen level inside the chamber, an oxygen analyzer was installed as shown in Figure 16. The oxygen analyzer was moved between the heating section, transfer section, and forming section to measure the changes in oxygen content over time, ensuring uniformity in the oxygen distribution. Figure 17 shows the history of oxygen content changes according to nitrogen gas pressure and time. It was confirmed that the oxygen content could be controlled to below 1% over time. However, when the nitrogen pressure was low at 5 kPa, it took over 80 min to control the oxygen content to below 1%, and non-uniformity in oxygen distribution across different positions was observed. To improve this, when the nitrogen gas pressure was increased to 7.2 kPa, the time required to achieve the same level of oxygen control was reduced to under 25 min, and the oxygen distribution became uniform across the chamber. Therefore, it is considered that a nitrogen gas pressure of 7.2 kPa is appropriate for atmosphere control.

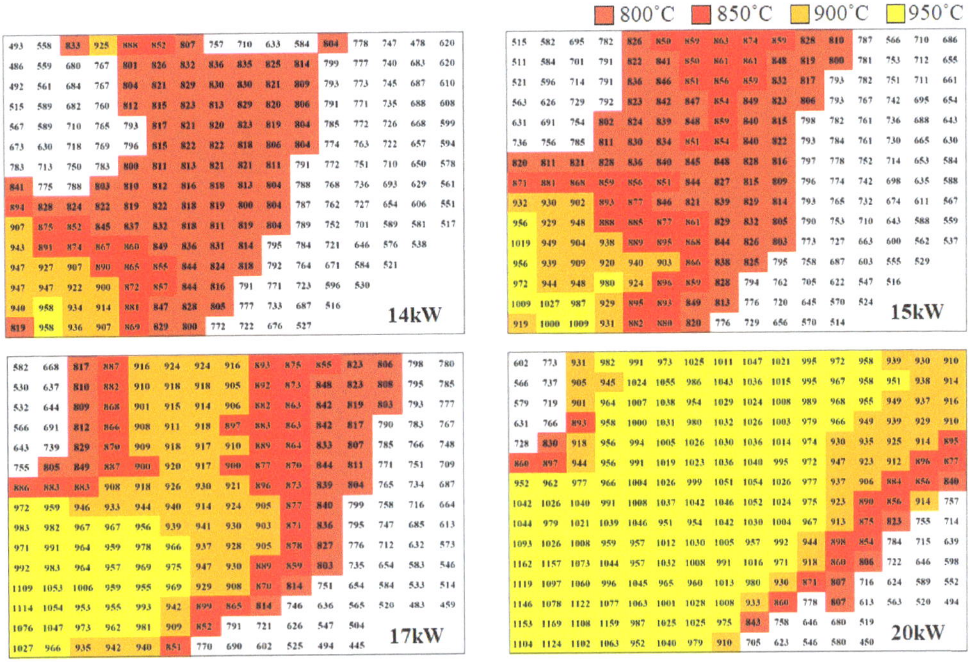

Figure 15. Temperature distribution according to the heating module output.

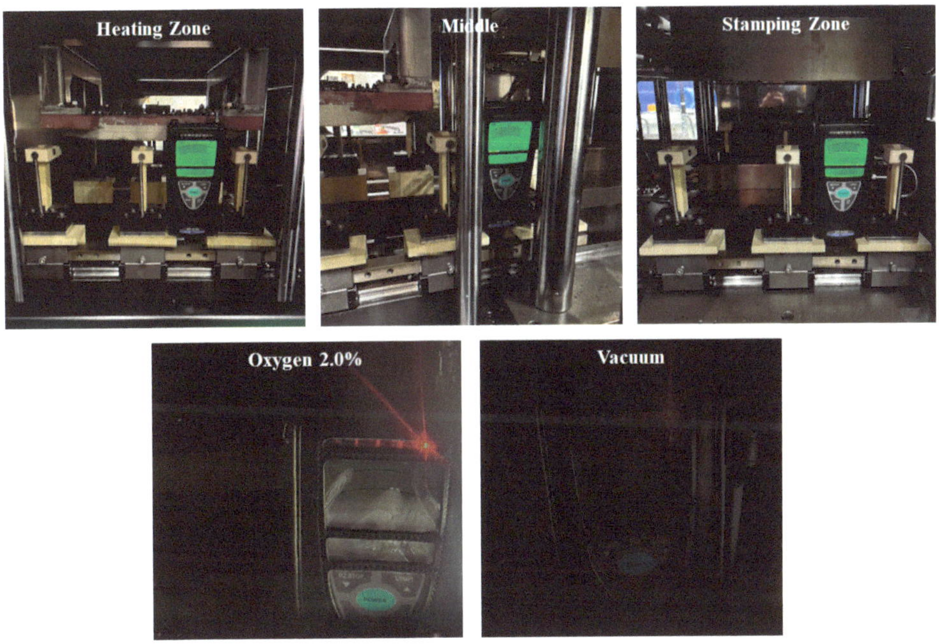

Figure 16. Installation of the oxygen analyzer inside the vacuum chamber.

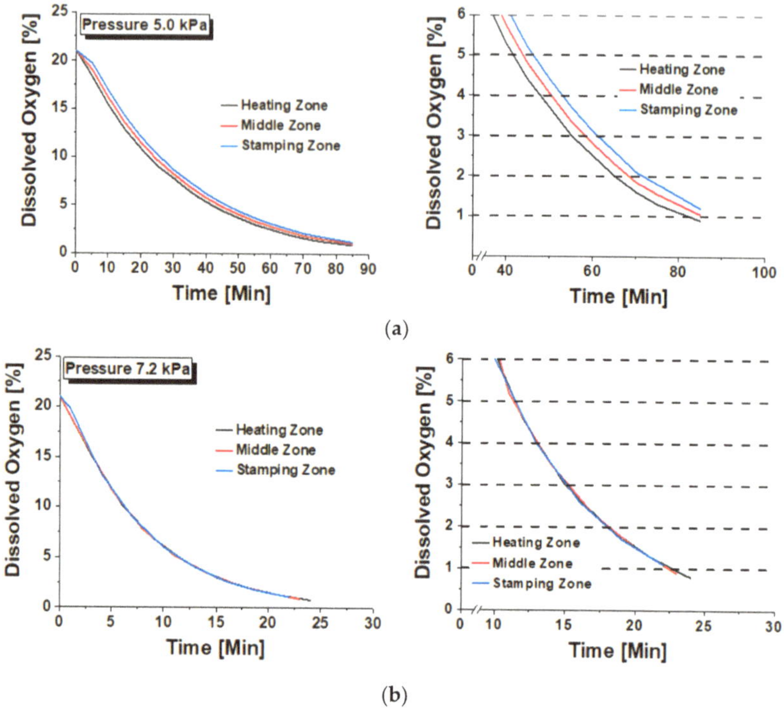

Figure 17. Atmosphere control test result according to the nitrogen gas pressure: (**a**) 5.0 kPa; (**b**) 7.2 kPa.

4.3. Evaluation of Oxide Layer

Forming tests on hat-type specimens made of uncoated steel sheets were conducted using a lab-scale hot stamping simulator, with heating and pressure control, to examine the effects of oxygen levels. After stabilizing the atmosphere according to the oxygen level, the material was heated by an induction coil, transferred to the hot stamping die by an automatic transfer device, and then formed into a hat-type specimen inside the die using water cooling. To observe the amount of oxide layer formation based on oxygen levels, forming tests were conducted under oxygen conditions of 0% (vacuum-like), 2%, 4%, 6%, and 21% (atmosphere), as shown in Figure 18. Under the vacuum-like condition, almost no oxide layer was observed. However, as the oxygen level increased, the amount of oxide layer also gradually increased, and at 6% oxygen, the oxide formation was almost similar to that observed under atmospheric conditions.

For a more quantitative observation, samples were taken from the top area under different oxygen conditions because the top area is the most susceptible to oxide layer formation. The thickness of the oxide layer was measured using an optical microscope. The equipment used for observations is the ECLIPSE MA22 (Nikon, Tokyo, Japan). Figure 19 shows the optical microscope images for each condition. When the amount of oxygen is close to a vacuum condition, almost no oxide layer forms. As the oxygen level gradually increases, the thickness of the oxide layer also increases proportionally, and at an oxygen level of 6%, an oxide layer similar to that formed under atmospheric conditions is observed. This shows that the oxide layer of uncoated boron steel is highly sensitive to the amount of oxygen.

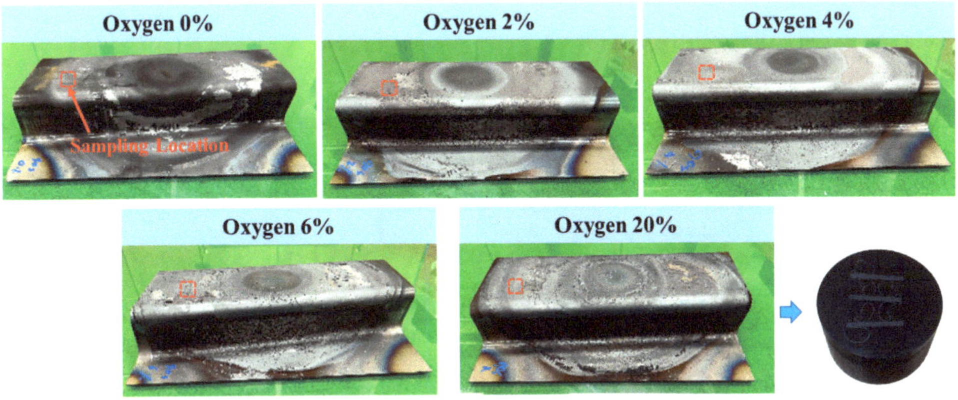

Figure 18. Experimental results of simulated specimens through atmosphere control.

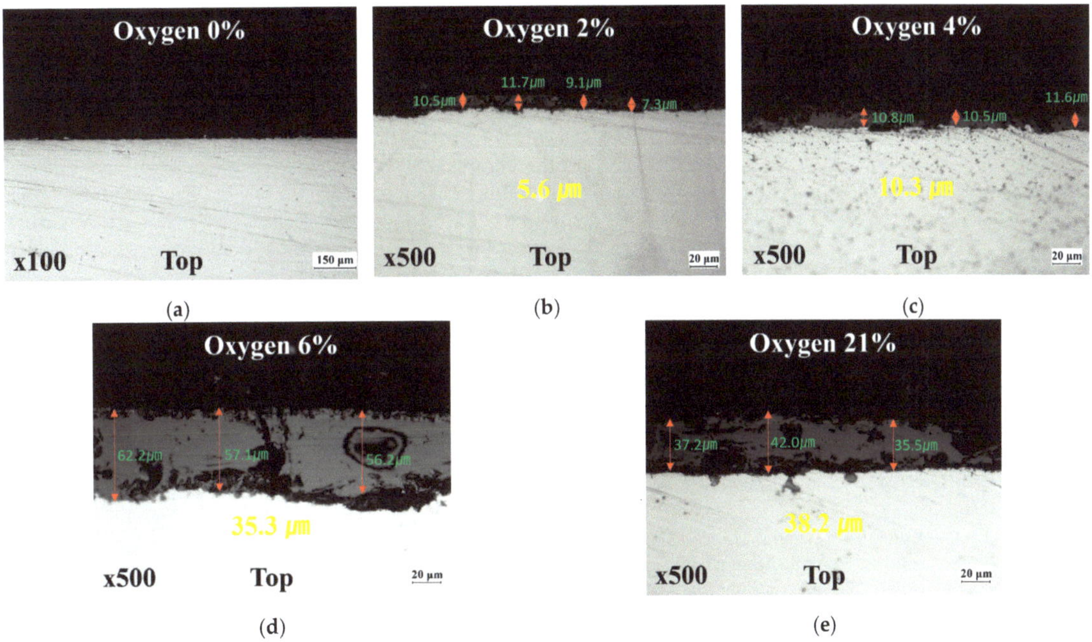

Figure 19. Optical microscope measurements according to atmosphere conditions at the top area: (**a**) oxygen 0%; (**b**) oxygen 2%; (**c**) oxygen 4%; (**d**) oxygen 6%; (**e**) oxygen 8%.

In the case of typical automotive parts, body assembly is carried out through processes such as spot welding after hot stamping. At this time, as the thickness of the residual oxide layer increases, the quality of welding can deteriorate rapidly. Therefore, for the hot stamping of uncoated boron steel, it is necessary to control the atmosphere to minimize the amount of oxygen in the environment during formation. Otherwise, surface oxide layer removal through processes such as shot peening will be required, which increases manufacturing costs and becomes an obstacle to using uncoated boron steel sheets. However, in actual component manufacturing processes, controlling the oxygen level close to a vacuum to suppress the oxide layer is highly challenging. It is more practical to introduce a threshold oxygen level condition in the hot stamping process of uncoated steel that ensures welding quality. Future research should focus on determining the optimal oxygen control conditions by analyzing the correlation between oxide layer formation and welding quality.

5. Conclusions

This study aimed to compare and analyze the surface property changes of boron steel during the hot stamping process, depending on the presence or absence of Al–Si coating, and to develop an atmosphere-controlled hot stamping technology for applying uncoated boron steel to ultra-high-strength automotive parts. Hot stamping experiments using a hat-shaped specimen showed that while no oxide layer formed on the Al–Si-coated steel, a thick oxide layer was observed on the uncoated steel due to exposure to atmospheric oxygen at high temperatures. This oxide layer not only deteriorates the surface quality of the uncoated steel but also significantly hinders weldability during the assembly of automotive parts. Surface hardness analysis revealed that the Vickers hardness of uncoated steel was, on average, 32 Hv higher than that of Al–Si-coated steel, suggesting that uncoated steel may be more advantageous for securing part strength. However, the formation of a thick oxide layer poses a significant risk to weld quality, making process technology to suppress oxide formation essential for the practical application of uncoated steel.

To address this issue, a new lab-scale hot stamping simulator was developed. This equipment was designed to perform forming experiments in an environment where the amount of oxygen in the atmosphere could be controlled to suppress oxide layer formation. The system uses a donut-shaped induction heating coil to heat the material to over 1000 °C and quickly transfer it to a forming die for rapid cooling and shaping through an automated transfer system. The oxygen concentration in the simulator was controlled using nitrogen gas and a vacuum pump, allowing the analysis of oxide layer formation under various oxygen levels. The experiment results showed that when the oxygen level in the atmosphere was below 6%, the thickness of the oxide layer decreased significantly, and in near-vacuum conditions, almost no oxide layer formed. These findings indicate that controlling the oxygen concentration can effectively suppress oxide layer formation in uncoated steel, improving both surface quality and weldability.

In conclusion, this study confirms that for uncoated boron steel to be applied in ultra-high-strength automotive parts, effective oxygen control during the hot stamping process is essential. By implementing such technology, it is expected that manufacturing costs can be reduced while simultaneously ensuring the strength and weld quality of uncoated steel parts.

Author Contributions: Conceptualization, J.L., G.B., and J.S.; methodology, G.B.; software, J.L.; validation, J.L. and G.B.; formal analysis, J.L.; investigation, G.B. and J.L.; resources, J.L.; data curation, J.L.; writing—original draft preparation, J.L. and G.B.; writing—review and editing, J.L. and G.B.; visualization, J.L. and G.B.; supervision, G.B. and J.S.; project administration, G.B. and J.S.; funding acquisition, G.B. and J.S. All authors have read and agreed to the published version of the manuscript.

Funding: This research was funded by the Korea Institute of Industrial Technology as "Development of intelligent root technology with add-on modules (KITECH EO-24-0005)" and by the Ministry of Trade, Industry and Energy (MOTIE, Korea) as "Development of multipoint servo control die cushion and process monitoring module for improving the press forming quality (20018327)" and "Development of Car Body Part Utilizing AlSi Coated Hot Press Forming Steel (20013403)".

Institutional Review Board Statement: Not applicable.

Informed Consent Statement: Not applicable.

Data Availability Statement: The original contributions presented in the study are included in the article, further inquiries can be directed to the corresponding author.

Conflicts of Interest: The authors declare no conflicts of interest.

References

1. Berglund, G. The History of Hardening of Boron Steel in Northern Sweden. In Proceedings of the 1st International Conference on Hot Sheet Metal Forming of High-Performance Steel, Kassel, Germany, 22–24 October 2008; pp. 175–177.
2. Aspacher, J. Forming Hardening Concepts. In Proceedings of the 1st International Conference on Hot Sheet Metal Forming of High Performance Steel, Kassel, Germany, 22–24 October 2008; pp. 77–81.
3. Merklein, M.; Lechler, J.; Stoehr, T. Characterization of Tribological and Thermal Properties of Metallic Coatings for Hot Stamping Boron-Manganese Steels. In Proceedings of the Seventh International Conference on Coatings in Manufacturing Engineering, 1–3 October 2008; pp. 1–3.
4. Naderi, M. Hot Stamping of Ultra High Strength Steels. Ph.D Thesis, Amirkabir University of Technology, Tehran, Iran, 2 November 2007.
5. Sato, H.; Watanabe, S. Microstructure and Hardness Development in Uncoated Steel Sheets After Hot Stamping. *Metall. Mater. Trans. B* **2017**, *48*, 2760–2772.
6. Bae, S.; Kang, Y. Effects of Cooling Rate on Mechanical Properties of Uncoated Hot Stamped Steels. *Steel Res. Int.* **2019**, *90*, 1900137.
7. Wang, C.; Liu, Q. Microstructural Analysis of Hot-Stamped Uncoated Alloys. *Metall. Mater. Trans. A* **2020**, *51*, 456–467.
8. Lee, J.; Park, H. Thermal and Structural Characteristics of Uncoated Steel Sheets During Hot Stamping. *Int. J. Mater. Form.* **2019**, *12*, 685–697.
9. Xu, F.; Zhang, Y. Experimental Investigation of Formability in Hot Stamping of Uncoated Sheets. *Int. J. Adv. Manuf. Technol.* **2020**, *106*, 1425–1432.
10. Liu, J.; Zhang, X. Optimization of Process Parameters for Uncoated Hot Stamping. *Materials* **2022**, *15*, 1357.
11. Altan, T.; Tekkaya, A.E. Analysis of Uncoated High-Strength Steels in Hot Stamping Applications. *J. Mater. Eng. Perform.* **2016**, *25*, 315–322.
12. Luo, X.; Dong, H. Comparative Study on Uncoated and Coated Steels in Hot Stamping Process. *J. Manuf. Process.* **2019**, *43*, 260–269.
13. Kim, S.; Kim, J.; Kim, Y. Advanced Hot Stamping Technology for Uncoated Materials. *Mater. Sci. Eng. A* **2018**, *734*, 85–93.
14. Li, X.; Wang, Z. Heat Transfer and Mechanical Behavior of Uncoated Sheets in Hot Stamping. *J. Manuf. Process.* **2017**, *29*, 251–259.
15. Gong, H.; Yang, S. Heat Transfer Modeling of Uncoated Steel Sheets in Hot Stamping Process. *Comput. Mater. Sci.* **2019**, *162*, 56–64.
16. Guo, J.; Wei, Y. Thermomechanical Behavior of Uncoated High-Strength Steels in Hot Stamping. *J. Mater. Process. Technol.* **2018**, *252*, 350–359.
17. Karbasian, H.; Tekkaya, A.E. A Review on Hot Stamping. *J. Mater. Process. Technol.* **2010**, *210*, 2103–2118. [CrossRef]
18. Borsetto, F.; Ghiotti, A.; Bruschi, S. Investigation of the High Strength Steel Al-Si Coating During Hot Stamping Operations. *Key Eng. Mater.* **2009**, *410*, 289–296. [CrossRef]
19. Mori, K.; Ito, D. Prevention of Oxidation in Hot Stamping of Quenchable Steel Sheet by Oxidation Preventive Oil. *CIRP Ann.* **2009**, *58*, 267–270 [CrossRef]
20. Chang, Y.; Tang, X.; Zhao, K.; Hu, P.; Wu, Y. Investigation of the Factors Influencing the Interfacial Heat Transfer Coefficient in Hot Stamping. *J. Mater. Process. Technol.* **2016**, *228*, 25–33. [CrossRef]
21. Taylor, T.; McCulloch, J. Effect of Part/Die Boundary Conditions on Microstructural Evolution During Hot Stamping 2000 MPa Class Boron Steel. *Steel Res. Int.* **2018**, *89*, 1700495. [CrossRef]
22. Chen, T.; Zhao, Y. Advantages and Limitations of Uncoated Hot Stamping in Automotive Industry Applications. *J. Mater. Process. Technol.* **2021**, *291*, 116–122.
23. Kim, Y.; Kang, M. Cost Efficiency of Using Uncoated Steels in Hot Stamping for Mass Production. *Procedia Manuf.* **2022**, *54*, 78–82.
24. Kobayashi, S.; Yakou, T. Control of Intermetallic Compound Layers at Interface Between Steel and Aluminum by Diffusion-Treatment. *Mater. Sci. Eng. A* **2002**, *338*, 44–53. [CrossRef]
25. Wang, K.; Gui, Z.; Liu, P.; Wang, Y.; Zhang, Y. Cracking Behavior of Al-Si Coating on Hot Stamping Boron Steel Sheet. *Procedia Eng.* **2014**, *81*, 1713–1718. [CrossRef]
26. Gui, Z.; Liang, W.; Liu, Y.; Zhang, Y. Thermo-Mechanical Behavior of the Al–Si Alloy Coated Hot Stamping Boron Steel. *Mater. Des.* **2014**, *60*, 26–33. [CrossRef]
27. Ghiotti, A.; Bruschi, S.; Borsetto, F. Tribological Characteristics of High Strength Steel Sheets Under Hot Stamping Conditions. *J. Mater. Process. Technol.* **2011**, *211*, 1694–1700. [CrossRef]
28. Kim, J.-H.; Ko, D.-C.; Lee, S.-B.; Kim, B.-M. Hardness Prediction in Hot Stamping Process by Local Blank Heating Based on Quench Factor Analysis. *Metals* **2018**, *9*, 29. [CrossRef]
29. Li, Y.; Chen, Y.; Li, S. Phase Transformation Testing and Modeling for Hot Stamping of Boron Steel Considering the Effect of the Prior Austenite Deformation. *Mater. Sci. Eng. A* **2021**, *821*, 141447. [CrossRef]

materials

MDPI

Review

Various Configurations for Improving the Efficiency of Metallic and Superconducting Photocathodes Prepared by Pulsed Laser Deposition: A Comparative Review

Alessio Perrone [1,2], Muhammad Rizwan Aziz [3,*] and Francisco Gontad [3]

1 Dipartimento di Matematica e Fisica "E. De Giorgi", Università del Salento, 73100 Lecce, Italy; alessio.perrone@unisalento.it
2 INFN-Istituto Nazionale di Fisica Nucleare, 73100 Lecce, Italy
3 Laser Applications Centre, Asociación de Investigación Metalúrgica del Noroeste, 36410 Porriño, Spain; francisco.gontad@aimen.es
* Correspondence: muhammadrizwan.aziz@unisalento.it

Abstract: This paper presents an innovative exploration of advanced configurations for enhancing the efficiency of metallic and superconducting photocathodes (MPs and SCPs) produced via pulsed laser deposition (PLD). These photocathodes are critical for driving next-generation free-electron lasers (FELs) and plasma-based accelerators, both of which demand electron sources with improved quantum efficiency (QE) and electrical properties. Our approach compares three distinct photocathode configurations, namely: conventional, hybrid, and non-conventional, focusing on recent innovations. Hybrid MPs integrate a thin, high-performance, photo-emissive film, often yttrium or magnesium, positioned centrally on the copper flange of the photo-injector. For hybrid SCPs, a thin film of lead is used, offering a higher quantum efficiency than niobium bulk. This study also introduces non-conventional configurations, such as yttrium and lead disks partially coated with copper and niobium films, respectively. These designs utilize the unique properties of each material to achieve enhanced photoemission and long-term stability. The novelty of this approach lies in leveraging the advantages of bulk photoemission materials like yttrium and lead, while maintaining the electrical compatibility and durability required for integration into RF cavities. The findings highlight the potential of these configurations to significantly outperform traditional photocathodes, offering higher QE and extended operational lifetimes. This comparative analysis provides new insights into the fabrication of high-efficiency photocathodes, setting the foundation for future advancements in electron source technologies.

Keywords: quantum efficiency; pulsed-laser deposition; thin films; metallic and superconducting photocathodes; photocathode configuration

Citation: Perrone, A.; Aziz, M.R.; Gontad, F. Various Configurations for Improving the Efficiency of Metallic and Superconducting Photocathodes Prepared by Pulsed Laser Deposition: A Comparative Review. *Materials* **2024**, *17*, 5257. https://doi.org/10.3390/ma17215257

Academic Editors: Guozheng Quan, Nikolaos Kladovasilakis, Konstantinos Tsongas and Dimitrios Tzetzis

Received: 24 July 2024
Revised: 30 September 2024
Accepted: 25 October 2024
Published: 29 October 2024

1. Introduction

Research into traditional metallic photocathodes (MPs) and superconducting photocathodes (SCPs) has attracted considerable attention due to their ability to generate high-brightness electron beams, their fast photoemission, which leads to a low temporal spread, the larger availability of electrons leading to the shortest time between electron pulses, or better electrical coupling with metallic cavities [1–5]. These technologies play crucial roles in the latest X-ray spectrum free-electron lasers (FELs) and are pivotal for advancements in particle-driven plasma wakefield accelerators [6–11]. The ongoing demand for electron beams characterized by high peak currents and minimal dark emittance has prompted researchers to explore alternative materials, moving beyond the conventional use of copper (Cu) and niobium (Nb) in Normal-Conducting Radio Frequency (NCRF) and Superconducting Radio Frequency (SRF) guns [12–16]. Recent years have seen progress in hybrid MPs, particularly those created by pulsed-laser deposition (PLD) on Cu substrates with yttrium (Y) and, to a lesser extent, magnesium (Mg) thin films. These metals have demonstrated higher quantum efficiency (QE) compared to Cu, yielding promising results [17–21]. Therefore, we have studied and developed new photocathode layouts called "hybrid configuration", based mainly on thin films of Mg or Y [18,21]. In the hybrid configuration depicted in Figure 1a, a layer of thin metal film is applied to the central area of the substrate using PLD. These films have shown strong adherence and a satisfactory photoemission performance, evidenced by their QE [17,22]. However, the laser cleaning process necessary before the photoemission effect reduces the thickness of these already thin films (0.1–2 μm), significantly shortening their lifespan [3,23,24].

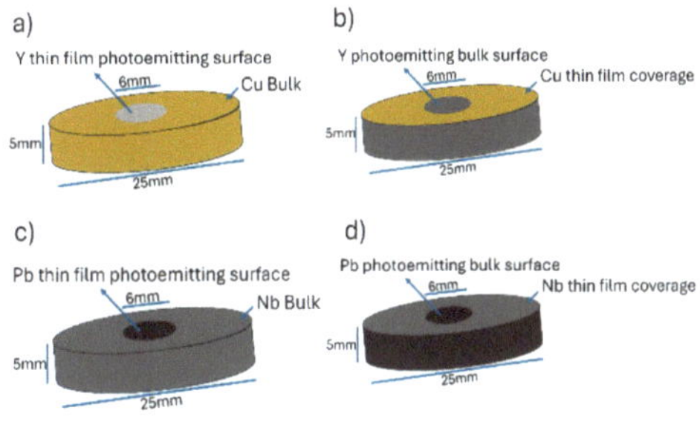

Figure 1. Scheme of the hybrid and non-conventional configurations for MP (**a**,**b**) and for SCP (**c**,**d**), respectively.

To address this issue, another novel configuration has been introduced [24]. Here, the thin film deposited by PLD is designed to improve the bulk material's photoemission and enhance the electrical compatibility with the RF-Gun cavity. This new MP prototype features a Y disk (25 mm diameter, 5 mm thickness) with a thin Cu film, leaving a 6 mm middle circular section of the Y substrate exposed as the emitting area, as shown in Figure 1b. This configuration allows the photoemission to occur directly from the Y bulk, while the RF gun cavity interacts with the surrounding Cu surface. This engineered setup integrates the advantageous combination of the high QE and low work function (WF) of Y with the necessary electrical compatibility suitable for Cu-based NCRF guns.

An analogous approach has been suggested for SCPs [25]. While pure Nb can function as a photocathode material within Nb SRF guns, its low QE limits the electron beam intensity [2]. Efforts to enhance SCP performance have included covering the Nb bulk with a thin lead (Pb) film [15], which is known as hybrid configurations which have also been explored, as illustrated in Figure 1c. However, Pb films tend to oxidize when exposed to air, which, although less severe than other semiconductor photocathodes, still poses a challenge [26,27]. Like MPs, the main drawback of hybrid SCPs is the thinning of materials due to the laser cleaning required to rejuvenate the emissive surface. To mitigate this, a similar innovative configuration has been proposed for SCPs. This involves a Pb substrate coated by a thin Nb film, with the central area masked to serve as the electron source (see Figure 1d). This design maintains the cavity's Q-value and does not compromise the applicable electric fields. These special configurations are not necessary to develop semiconductor photocathodes, such as gallium arsenide (GaAs) and cesium telluride (Cs_2Te) for their superior QE compared to traditional metal photocathodes. GaAs, in particular, is notable for its negative electron affinity surface, which dramatically enhances QE when activated with cesium and oxygen [28,29]. However, these materials often suffer from rapid degradation due to chemical reactions with residual gases in the vacuum system, which necessitates frequent reactivation [30]. Furthermore, the operational lifespan of semiconductor photocathodes can be limited by the damage incurred during the high-intensity laser cleaning process, similar to the challenges faced with metallic and superconducting counterparts [30]. Advanced techniques, such as atomic layer deposition (ALD) and molecular beam epitaxy (MBE), are being explored to create more robust and long-lasting semiconductor photocathodes by precisely controlling the film thickness and composition.

In this paper, we present our findings on MPs and SCPs, focusing on the non-conventional configurations developed to enhance the photoemission performance and extend their operational lifespan. The advantages and disadvantages of the three different photocathode configurations are reported in Table 1.

Table 1. Advantages and disadvantages of the different photocathode configurations.

Configuration	MP	SCP	Lifetime	QE
Conventional	Cu disk	Nb disk	Relatively long [1]	Relatively small [2]
Hybrid	Cu disk with a Y thin film deposited on the central area	Nb disk with a Pb thin film deposited on the central area	Shorter than conventional [3]	Higher than conventional [4]
Non-conventional	Y disk with a Cu annular thin film	Pb disk with a Nb annular thin film	Much higher than hybrid [5]	Higher than conventional [4]

[1] High chemical inertness; [2] High WF; [3] Thinning of the film due to the laser cleaning treatment; [4] Low WF; [5] Bulk as photoelectron source.

2. The Choice of the Deposition Technique

Various techniques have been employed to deposit metallic thin films used as electron sources [15,31,32]. Among these methods, the PLD technique stands out for its ability to ensure the robust bonding of metallic films to substrates, even when applied at an ambient temperature [21,32,33]. When selecting a deposition technique for metals such as Cu, Mg, and Pb for photocathode applications, the PLD technique presents distinct advantages over other Physical Vapor Deposition (PVD) methods. Techniques such as thermal evaporation, electron beam evaporation, sputter deposition, cathodic arc deposition, and molecular beam epitaxy (MBE) each have specific drawbacks that make them less suitable for high-quality photocathode fabrication.

Thermal evaporation often suffers from contamination due to the inclusion of impurities in the evaporated material. Additionally, achieving precise control over the film thickness is challenging, and scaling the process for larger substrates is problematic [34]. Moreover, the adherence of the film deposited by this technique is poor [31]. Electron beam evaporation, while offering some improvements, introduces high equipment costs and substrate heating issues that can damage sensitive materials. This method also struggles with precise thickness control, particularly for materials with high melting points [35].

Sputter deposition, another common PVD technique, frequently induces stress in the films, leading to potential cracking or delamination. Its lower deposition rates extend the process times, and the high-energy ions used can damage the substrate [36]. Cathodic arc deposition, although capable of high deposition rates, produces droplets that result in rough surfaces and film defects. This method also causes significant substrate heating and potential contamination from the cathode material [34].

MBE, known for producing high-quality films, is hindered by extremely high costs, slow deposition rates, and the requirement for ultra-high vacuum conditions. These factors increase the operational complexity and restrict the throughput, making MBE less suitable for large-scale applications [37].

In contrast, PLD offers numerous advantages. It can produce high-quality, uniform thin films with smooth surfaces and provides precise control of the deposited films' thickness. PLD is versatile, capable of depositing a wide range of materials, including complex alloys and compounds, while minimizing contamination due to the laser ablation process. Despite its relatively low deposition rates, it is suitable for thin films, and can be optimized to minimize substrate heating and damage. These attributes make PLD a superior choice for the deposition of metals like Cu, Mg, and Pb for photocathode applications [38].

A common setup for PLD involves a high vacuum chamber made of stainless steel (10^{-5}–10^{-6} Pa), where the substrate is placed parallel to the target, typically 4–5 cm apart.

To prevent the development of significant craters or directional formations, the target undergoes rotation and displacement during deposition, a practice followed in our previous research. Ultraviolet-visible lasers are frequently utilized due to their effective interaction with various target materials. The pulsed laser beam strikes the target surface at a 45° angle, delivering a power density ranging from 0.1 to 1 GW/cm^2, within nanosecond durations. Depending on specific parameters, metallic films are deposited on the substrate at rates typically measured in tenths of an angstrom per laser pulse. A notable strength of PLD lies in the strong adherence of these films to the substrate, a characteristic that has been consistently highlighted. Measurements using time-of-flight techniques indicate that the plasma's average ion kinetic energy can vary significantly, potentially reaching several hundred electron volts, depending on the laser intensity [39–42]. The QE of the films is typically measured in a high vacuum system where the photocathode is aligned in front of the anode such that a laser beam can be focused on the photocathode surface without affecting the anode. An intense electric field, within the megavolt per meter range, is applied between the anode and cathode, and the emitted current is measured with respect to the laser energy hitting the photocathode surface. The QE of the photocathodes prepared by PLD was studied using a custom-built photodiode cell, designed according to the specifications mentioned earlier, as illustrated in Ref. [31].

3. Photocathodes Assembly

3.1. Hybrid and Non-Conventional Configurations of Metallic Photocathodes

Previous studies extensively cover the operational principles and characteristics of traditional MPs [3,19,43]. Copper is frequently used in these applications due to its high chemical stability, long lifespan, low electrical resistivity, good thermal conductivity and economical cost, which enhance the quality factor of RF cavities. Nevertheless, Cu exhibits a low QE of 2.0×10^{-5} at 266 nm [17], necessitating the utilization of powerful UV lasers to overcome its high WF (see Table 2). Even photocathodes based on Mg with a WF of 3.6 eV require the use of the fourth harmonic of a Nd:YAG laser (266 nm, 4.66 eV) or the third harmonic of a Ti:sapphire (267 nm, 4.64 eV). That means much less energy and stability of the drive laser. On the contrary, the low WF value of Y (3.1 eV) allows one to drive photocathodes based on this metal with the third harmonic (355 nm, 3.5 eV) of the Nd:YAG laser. However, many facilities still use Mg [44] as a photocathode due to its relatively high QE (7.6×10^{-4} at 266 nm) and to the lower chemical reactivity than that of Y. These two metals react with hydrogen- and oxygen-containing molecules such as H_2, H_2O, CO_2, and O_2, forming hydrides and oxides on the surface [45]. However, it is important to note that the oxidation processes in bulk metal photocathodes are significantly less severe compared to those in semiconductor photocathodes. Recently, "smart" cathodes designed with thin films to enhance QE and extend longevity have been developed. The hybrid design, featuring a thin film realized on the middle part of a conventional metallic cathode (see Figure 1a,c), has been implemented for both MPs and SCPs. Figure 2a,b illustrate the collected charge vs. the function of laser energy for Y thin films on Cu in a hybrid configuration and for Cu bulk after laser cleaning. The linear correlation between the collected charge and laser energy suggests that photoemission takes place through a one-photon absorption mechanism. Additionally, the minimal space charge effect, resulting from the low collected charge, further supports this. The slope corresponding to the continuous line correlates to the QE value [32]. Figure 2a,b illustrate that the QE of the Y thin film on Cu is substantially greater than that of bulk Cu under identical conditions and is also corroborated by the literature [21]. The most important photo-emissive properties of the metals used in metallic and superconducting photocathodes are shown in Table 2.

Table 2. Relevant photo-emissive properties of the most used bulk metallic photocathodes (conventional configuration).

	Cu	Mg	Y	Nb	Pb
Work function (eV)	4.6 [46]	3.6 [47]	3.1 [48]	4.3 [49]	4.2 [49]
QE (at 266 nm) Bulk	2.0×10^{-5} [17]	7.6×10^{-4} [21]	3.0×10^{-4} [21]	3.2×10^{-6} at 248 nm [13]	7×10^{-5} [50]
Operational lifetime (yrs) [51]	No limit	~1	<1	No limit	>1
Compatibility with a Cu RF gun	Very high	Medium	Low	-	-
Compatibility with a Nb SCRF gun	-	-	-	Very high	High

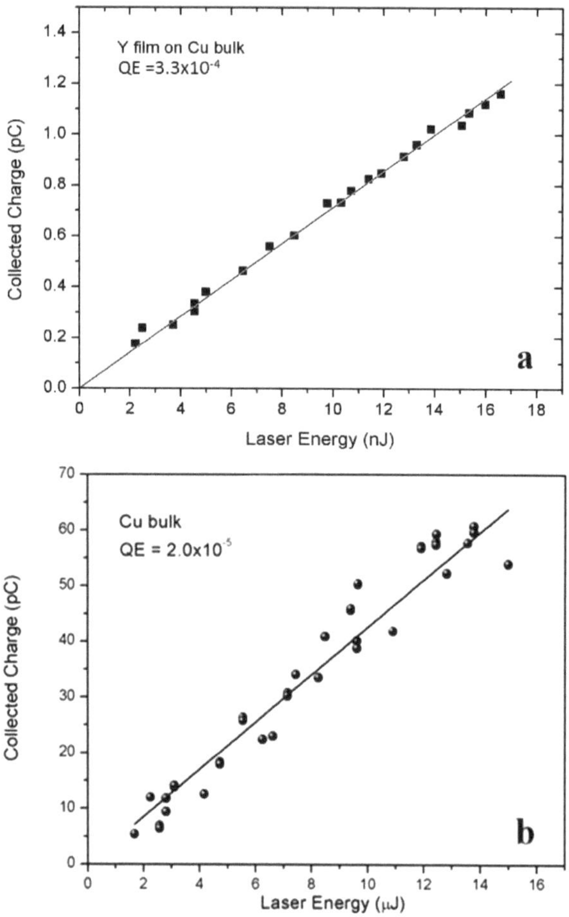

Figure 2. Collected charge as a function of laser energy after laser cleaning treatment for Y thin film deposited on Cu in hybrid configuration (**a**) and for Cu bulk (**b**). Continuous lines are the data-fitting curves (Figure 2a reproduced from ref. [21] and Figure 2b reproduced from ref. [17]).

As previously noted, Y and Mg films deposited by the PLD technique typically range from hundreds of nanometers to a few micrometers in thickness, often insufficient to endure repeated cleaning processes required for operational use. Thicker films, whether deposited by PLD or other methods, often result in high mechanical strain and cracking, leading to the detachment of the film from the RF gun back flange. Several approaches have been suggested to resolve this problem, including friction welding [52], hot isostatic pressing [53] and press-fitting bulk disks [47].

However, inserting a disk of another material into the middle section of a Cu photocathode introduces minute discontinuities between the two metals, causing undesirable electrical discharges and reducing the RF cavity's quality factor. On the contrary, the discontinuity between the substrate and the thin film in the non-conventional layout is very smooth. Thus, the non-conventional configuration mentioned earlier and shown in Figure 1b,d appears to be the most effective alternative to prevent these issues and the thinning process [48]. This innovative design allows the metallic photocathode to retain the photoemissive properties of Y bulk [24]. Indeed, with the present assembly, the laser cleaning process is related to bulk metal (Y or Mg) and not to the thin film. The most thoroughly investigated MP in this non-conventional setup is a Cu thin film annularly realized on a Y disk (see Figure 1b), combining photoemissive characteristics of Y with electrical properties of Cu. The conductive Cu thin films were analyzed morphologically, structurally and electrically to ensure their suitability for application. Scanning electron microscopy (SEM) showed the presence of small droplets on the Cu surface, albeit in smaller quantities compared to other metal deposition techniques [2].

The structural characteristics of the Cu film on the Y substrate, deposited by sub-ps laser pulses, utilized as a photocathode, were analyzed by X-ray diffraction (XRD). The XRD pattern in Figure 3 indicates peaks corresponding to the (111), (200) and (220) crystallographic planes of Cu, while other peaks are attributed to the Y polycrystalline substrate. Both morphological and structural analyses indicate the Cu film's promising properties for application, and its high adherence to the Y substrate is also confirmed.

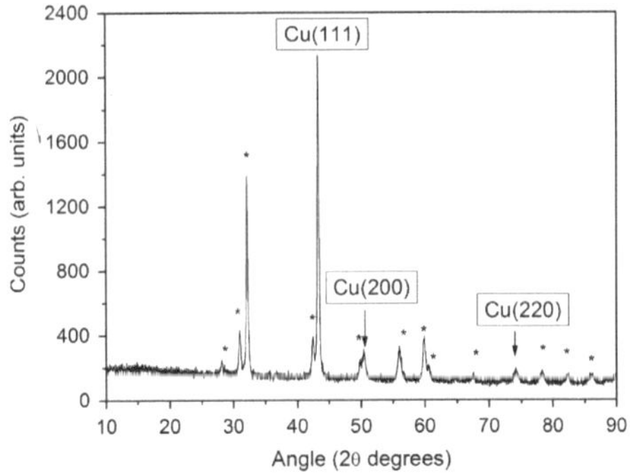

Figure 3. XRD pattern of Cu film deposited by PLD technique on a Y disk by sub-ps laser pulses. The peaks labeled as * can be ascribed to the Y polycrystalline (reproduced from ref. [48]).

This excellent physical performance is mainly tied to the substantial kinetic energy of particles within the Cu plasma during the laser ablation procedure. The mean electrical resistivity of a 100 nm thick Cu film on a Y substrate was measured at 2.3×10^{-6} Ω·cm, translating to a sheet resistance of 3.1×10^{-1} Ω/sq [24]. This measured resistivity value is about 36% higher than that of bulk high-purity Cu (1.69×10^{-6} Ω × cm) [54], likely due to electron scattering caused by the atomic-scale surface roughness of the deposited film [55].

3.2. *Hybrid and Non-Conventional Configurations of Superconducting Photocathodes*

Superconducting photocathodes leverage the unique properties of superconducting materials to generate high-brightness electron beams with minimal energy dissipation. Nb and Pb are the most commonly used materials in SCPs due to their favorable superconducting properties. Due to its high critical temperature and good mechanical properties, is widely used in SRF cavities. Pb, although having a lower critical temperature, offers a higher QE when used as a photoemissive material. The combination of these materials in hybrid configurations aims to utilize the high QE of Pb and the robust superconducting characteristics of Nb [25]. The deposition methods, like PLD, for creating these hybrid configurations are critical as they affect the film's adherence, uniformity, and overall performance. The challenge of maintaining surface cleanliness and preventing oxidation is paramount, as these factors directly impact the photoemissive efficiency and lifespan of SCPs. The hybrid configuration for superconducting photocathodes (SCPs) primarily incorporates Nb and Pb metals. These devices are created by depositing a thin Pb film onto a Nb substrate employing numerous deposition methods, such as electroplating, arc deposition, sputtering, and thermal evaporation [15,31,34,56–58]. Figure 4 shows the XRD pattern of the Nb film, indicating that the PLD film is largely amorphous, with a weak peak around 38° corresponding to the (110) crystalline planes. Other peaks are attributed to the Pb substrate.

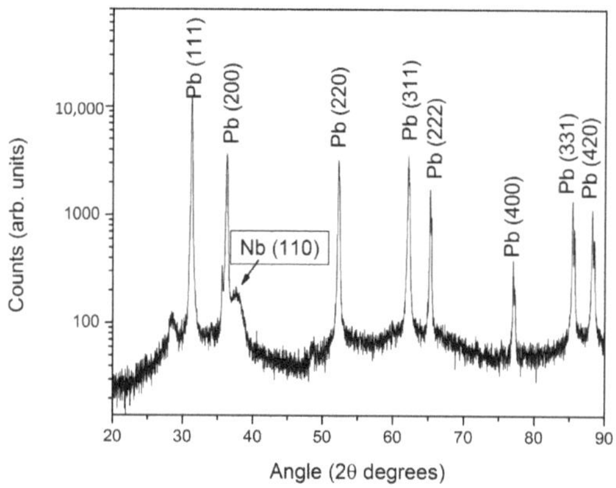

Figure 4. XRD pattern of the Nb PLD film deposited on Pb substrate (reproduced from ref. [25]).

Figure 5a,b display the electron charge emitted from a Pb thin film prepared by PLD in a hybrid configuration compared to Pb bulk versus the laser energy.

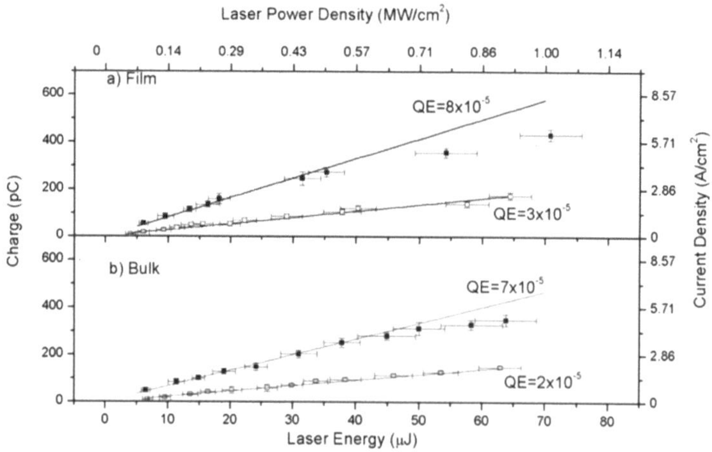

Figure 5. Charge of the electrons emitted from (**a**) Pb film prepared by PLD in hybrid configuration and (**b**) Pb bulk before (□) and after (■) laser cleaning treatment (reproduced from ref. [50]).

The QE values for Pb bulk and Pb thin film are 7×10^{-5} and 8×10^{-5} at 266 nm, respectively [50], which are significantly higher than the 3.2×10^{-6} QE of the Nb bulk at 248 nm [13]. However, it is observed that a slight space charge effect occurs at laser energies exceeding 35 μJ. Despite the good chemical stability of Pb, its surface oxidizes when exposed to air, necessitating laser cleaning before its use in SRF guns. As with MPs, these cleaning processes lead to the thinning of the photoemitting surface, compromising the performance and lifespan. Consequently, alternative mechanical and physical solutions have been explored by various research groups.

In a non-conventional set-up, a configuration was proposed where a Pb disk is entirely coated with a Nb thin film, excluding its central surface, which serves as the photo-emitting area (Figure 1d) [25]. As in the case of metallic photocathodes, in the non-conventional configuration the drive laser acts on the Pb bulk and not on the thin film. Mechanical solutions face similar challenges as those with MPs; thus, PLD appears to be an optimal technique for depositing highly adherent thin films. The Nb films were analyzed for their morphological, structural and electrical properties, as well as mechanical properties.

Nano-indentation analysis determined the mechanical properties of the Nb film, revealing a good hardness value of 2.8 ± 0.3 GPa [25]. It is crucial to avoid surface scratches on photocathodes, as these can cause electrical discharges under the high electric fields in superconducting cavities.

Another vital characteristic of SCPs is their electrical compatibility with superconducting cavities. Figure 6 shows the sample resistance as a function of temperature, with two noticeable discontinuities: one at 9.3 K, corresponding to the critical temperature of Nb, and another at 7.3 K, near the critical temperature of Pb [25].

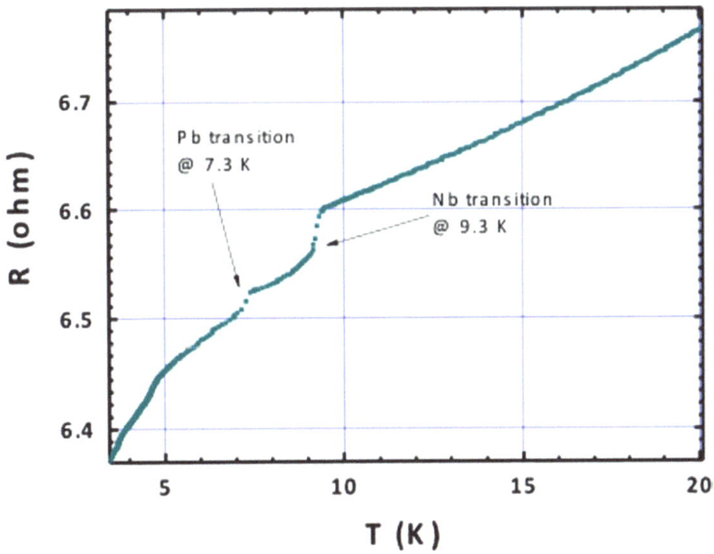

Figure 6. Electrical resistance evolution of the sample Nb thin film on Pb substrate with the temperature (reproduced from ref. [25]).

In this non-conventional configuration, SCPs have similar photoelectrical properties to Pb bulk. These findings indicate that the non-conventional setups for both MPs and SCPs show the greatest potential and dependability among the various proposed configurations (see Tables 1–4).

Table 3. Photo-emissive properties of PLD thin film in hybrid configuration.

	Mg	**Y**	**Pb**
QE (at 266 nm)	1.8×10^{-3} [21]	3.3×10^{-4} [21]	8.0×10^{-5} [50]
Operational lifetime *	2–3 months [18]	3–5 months *	4–6 months [31]
Compatibility with a Cu RF gun	High	Medium	-
Compatibility with a Nb SCRF gun	-	-	High

* Expected value.

Table 4. Photo-emissive properties of PLD thin film in non-conventional configuration.

	Mg Bulk Cathode with Cu-Mask	**Y Bulk Cathode with Cu-Mask**	**Pb Bulk Cathode with Nb-Mask**
QE (at 266 nm)	7.6×10^{-4}	3×10^{-4}	7×10^{-5} [50]
Operational lifetime (yrs) Similar to those of bulk materials given in Table 2	~1	<1	>1
Compatibility with a Cu RF gun	Very high	High	-
Compatibility with a Nb SCRF gun	-	-	Very high

4. Conclusions

This study presents a comprehensive exploration of advanced photocathode configurations aimed at enhancing the performance of MPs and SCPs fabricated via PLD. The comparative analysis of conventional, hybrid, and non-conventional configurations demonstrates significant improvements in QE, operational longevity, and integration with RF cavities.

The hybrid configuration, where thin films of high-performance materials such as Y or Pb are deposited onto conventional Cu or Nb substrates, offers distinct advantages. This approach enhances the photoemission properties while maintaining good electrical compatibility with RF cavities. The thin film design allows for a higher QE than conventional metallic photocathodes, making it a viable solution for improving the efficiency of electron sources. However, the hybrid configuration faces challenges related to the thinning of the photoemissive layer during the laser cleaning process, which limits its operational lifespan. Despite this, the hybrid photocathodes represent a significant step forward in leveraging thin-film technology to optimize the photocathode performance.

In parallel, the non-conventional configuration represents a further advancement by mitigating the thinning issue. By using bulk Y and Pb materials directly exposed to the drive laser, while surrounding these with Cu or Nb films for electrical compatibility, this configuration combines the best of both worlds: the superior photoemission of Y and Pb and the robust structural and electrical properties of Cu and Nb. This design not only extends the operational lifespan, but also maintains a high QE over time, positioning the non-conventional configuration as a highly durable and efficient option for demanding applications. Both the hybrid and non-conventional configurations underscore the novelty of this study, as they provide new pathways for improving the photoemission efficiency and stability of photocathodes used in next-generation accelerators such as FELs and plasma wakefield accelerators. The hybrid configuration offers a near-term solution with a higher QE than conventional designs, while the non-conventional configuration pushes the boundaries of photocathode technology by addressing long-term operational issues.

These findings lay the groundwork for future innovations in photocathode design, demonstrating the potential for integrating advanced materials and configurations to enhance the electron source performance. The contributions of both configurations represent substantial progress toward meeting the stringent requirements of modern accelerator technologies.

Author Contributions: Conceptualization and writing—original draft preparation, A.P.; formal analysis, writing—review and editing, F.G.; validation and supervision, M.R.A. All authors have read and agreed to the published version of the manuscript.

Funding: This research received no external funding.

Acknowledgments: This work was supported by the Italian National Institute of Nuclear Physics (INFN). M. R. Aziz would like to express sincere gratitude to the Center of Applied Physics, Dating and Diagnostics (CEDAD) for the unwavering support to this research.

Conflicts of Interest: The authors declare no conflict of interest.

References

1. Rosenzweig, B.; Cahill, A.; Dolgashev, V.; Emma, C.; Fukasawa, A.; Li, R.; Limborg, C.; Maxson, J.; Musumeci, P.; Nause, A.; et al. Next generation high brightness electron beams from ultrahigh field cryogenic rf photocathode sources. *Phys. Rev. Accel. Beams* **2019**, *22*, 023403. [CrossRef]
2. Hannon, F.; Andonian, G.; Harris, L.A. A Plasmonic Niobium Photocathode for Srf Gun Applications. In Proceedings of the 10th International Particle Accelerator Conference IPAC2019, Melbourne, Australia, 19–24 May 2019; pp. 19–24.
3. Zhou, F.; Brachmann, A.; Decker, F.J.; Emma, P.; Gilevich, S.; Iverson, R.; Stefan, P.; Turner, J. High-Brightness Electron Beam Evolution Following Laser-Based Cleaning of a Photocathode. *Phys. Rev. Spec. Top.-Accel. Beams* **2012**, *15*, 090703. [CrossRef]
4. Faillace, L.; Agustsson, R.; Behtouei, M.; Bosco, F.; Bruhwiler, D.; Camacho, O.; Carillo, M.; Fukasawa, A.; Gadjev, I.; Giribono, A. High Field Hybrid Photoinjector Electron Source for Advanced Light Source Applications. *Phys. Rev. Accel. Beams* **2022**, *25*, 063401. [CrossRef]
5. Xu, X.; Li, F.; Tsung, F.S.; Miller, K.; Yakimenko, V.; Hogan, M.J.; Joshi, C.; Mori, W.B. Generation of ultrahigh-brightness pre-bunched beams from a plasma cathode for X-ray free-electron lasers. *Nat. Commun.* **2022**, *13*, 3364. [CrossRef]
6. Decking, W.; Abeghyan, S.; Abramian, P.; Abramsky, A.; Aguirre, A.; Albrecht, C.; Alou, P.; Altarelli, M.; Altmann, P.; Amyan, K. A MHz-Repetition-Rate Hard X-Ray Free-Electron Laser Driven by a Superconducting Linear Accelerator. *Nat. Photonics* **2020**, *14*, 391–397. [CrossRef]
7. Filippetto, D.; Grames, J.; Hernandez-Garcia, C.; Karkare, S.; Piot, P.; Power, J.; Sun, Y.; Wang, E. Electron Sources for Accelerators. *arXiv* **2022**, arXiv:2207.08875.
8. Massimo, F.; Marocchino, A.; Chiadroni, E.; Ferrario, M.; Mostacci, A.; Musumeci, P.; Palumbo, L. Transformer Ratio Studies for Single Bunch Plasma Wakefield Acceleration. *Nucl. Instrum. Methods Phys. Res. A* **2014**, *740*, 242–245. [CrossRef]
9. Nie, Z.; Li, F.; Morales, F.; Patchkovskii, S.; Smirnova, O.; An, W.; Zhang, C.; Wu, Y.; Nambu, N.; Matteo, D. Highly Spin-Polarized Multi-GeV Electron Beams Generated by Single-Species Plasma Photocathodes. *Phys. Rev. Res.* **2022**, *4*, 033015. [CrossRef]
10. Li, R.K.; To, H.; Andonian, G.; Feng, J.; Polyakov, A.; Scoby, C.M.; Thompson, K.; Wan, W.; Padmore, H.A.; Musumeci, P. Surface-Plasmon Resonance-Enhanced Multiphoton Emission of High-Brightness Electron Beams from a Nanostructured Copper Cathode. *Phys. Rev. Lett.* **2013**, *110*, 074801. [CrossRef]
11. Hidding, B.; Assmann, R.; Bussmann, M.; Campbell, D.; Chang, Y.-Y.; Corde, S.; Cabadağ, J.C.; Debus, A.; Döpp, A.; Gilljohann, M.; et al. Progress in Hybrid Plasma Wakefield Acceleration. *Photonics* **2023**, *10*, 99. [CrossRef]
12. Barday, R.; Kamps, T.; Kugeler, O.; Neumann, A.; Schmeißer, M.; Voelker, J.; Smedley, J.; Sekutowicz, J.; Kneisel, P.; Nietubyc, R. Characterization of a Superconducting Pb Photocathode in a SRF Gun Cavity. In Proceedings of the IPAC2013, Shanghai, China, 12–17 May 2013; pp. 279–281.
13. Hand, L.N.; Happek, U. Photoelectric quantum efficiency of niobium for λ = 193 nm and λ = 248 nm. *Nucl. Instrum. Methods Phys. Res. A* **1996**, *372*, 335–338. [CrossRef]
14. Qian, H.J.; Murphy, J.B.; Shen, Y.; Tang, C.X.; Wang, X.J. Surface Photoemission in a High-Brightness Electron Beam Radio Frequency Gun. *Appl. Phys. Lett.* **2010**, *97*, 253504. [CrossRef]
15. Smedley, J.; Rao, T.; Sekutowicz, J. Lead Photocathodes. *Phys. Rev. Spec. Top.-Accel. Beams* **2008**, *11*, 013502. [CrossRef]
16. Németh, K.; Harkay, K.C.; van Veenendaal, M.; Spentzouris, L.; White, M.; Attenkofer, K.; Srajer, G. High-Brightness Photocathodes through Ultrathin Surface Layers on Metals. *Phys. Rev. Lett.* **2010**, *104*, 046801. [CrossRef]
17. Lorusso, A.; Trovò, M.; Demidovich, A.; Cinquegrana, P.; Gontad, F.; Broitman, E.; Chiadroni, E.; Perrone, A. Pulsed Laser Deposition of Yttrium Photocathode Suitable for Use in Radio-Frequency Guns. *Appl. Phys. A* **2017**, *123*, 779. [CrossRef]
18. Cultrera, L.; Gatti, G.; Miglietta, P.; Tazzioli, F.; Perrone, A.; Moody, J.T.; Musumeci, P. Electron Emission Characterization of Mg Photocathode Grown by Pulsed Laser Deposition within an S-Band Rf Gun. *Phys. Rev. Spec. Top.-Accel. Beams* **2009**, *12*, 043502. [CrossRef]
19. Scifo, J.; Alesini, D.; Anania, M.P.; Bellaveglia, M.; Bellucci, S.; Biagioni, A.; Bisesto, F.; Cardelli, F.; Chiadroni, E.; Cianchi, A. Nano-Machining, Surface Analysis and Emittance Measurements of a Copper Photocathode at SPARC_LAB. *Nucl. Instrum. Methods Phys. Res. A* **2018**, *909*, 233–238. [CrossRef]
20. Henneken, H.; Scholze, F.; Krumrey, M.; Ulm, G. Quantum efficiencies of gold and copper photocathodes in the VUV and X-ray range. *Metrologia* **2000**, *37*, 485. [CrossRef]

21. Lorusso, A.; Gontad, F.; Solombrino, L.; Chiadroni, E.; Broitman, E.; Perrone, A. Tight Comparison of Mg and Y Thin Film Photocathodes Obtained by the Pulsed Laser Deposition Technique. *Nucl. Instrum. Methods Phys. Res. A* **2016**, *836*, 57–60. [CrossRef]
22. Xiang, R.; Arnold, A.; Michel, P.; Murcek, P.; Teichert, J.; Lu, P.; Vennekate, H.; Patra, P. Improvement of photoemission efficiency of magnesium photocathodes. In Proceedings of the IPAC, Copenhagen, Denmark, 14–19 May 2017.
23. Siewert, F.; Smedley, J.; Weinberg, G.; Barday, R.; Kamps, T.; Quast, T.; Schubert, S.G.; Varykhalov, A.; Nietubyc, R. Investigation of Laser-cleaning Process on Lead Photocathodes. In Proceedings of the 3rd International Conference on Particle Accelerator (IPAC 2012), New Orleans, LA, USA, 20–25 May 2012.
24. Perrone, A.; D'Elia, M.; Gontad, F.; Di Giulio, M.; Maruccio, G.; Cola, A.; Stankova, N.E.; Kovacheva, D.G.; Broitman, E. Non-Conventional Photocathodes Based on Cu Thin Films Deposited on Y Substrate by Sputtering. *Nucl. Instrum. Methods Phys. Res. A* **2014**, *752*, 27–32. [CrossRef]
25. Gontad, F.; Lorusso, A.; Panareo, M.; Monteduro, A.G.; Maruccio, G.; Broitman, E.; Perrone, A. Nanomechanical and Electrical Properties of Nb Thin Films Deposited on Pb Substrates by Pulsed Laser Deposition as a New Concept Photocathode for Superconductor Cavities. *Nucl. Instrum. Methods Phys. Res. A* **2015**, *804*, 132–136. [CrossRef]
26. Bellani, S.; Najafi, L.; Martín-García, B.; Ansaldo, A.; Del Rio Castillo, A.E.; Prato, M.; Moreels, I.; Bonaccorso, F. Graphene-Based Hole-Selective Layers for High-Efficiency, Solution-Processed, Large-Area, Flexible, Hydrogen-Evolving Organic Photocathodes. *J. Phys. Chem. C* **2017**, *121*, 21887–21903. [CrossRef]
27. Mine, P.; Malamud, G.; Vartsky, D.; Vasileiadis, G. Organic UV photocathodes for gaseous particle detectors. *Nucl. Instr. Meth.* **1997**, *387*, 171–175. [CrossRef]
28. Bae, J.K.; Bazarov, I.; Cultrera, L.; Galdi, A.; Ikponmwen, F.; Maxson, J. Enhanced Robustness of GaAs-Based Photocathodes Activation by Cs, Sb, and O_2. In Proceedings of the North American Particle Accelerator Conference (NAPAC'19), Lansing, MI, USA, 1–6 September 2019; JACOW Publishing: Geneva, Switzerland, 2019; pp. 210–212.
29. Liu, Z.; Guan, J.; Dai, Z.; Zhong, J.; Liu, L.; Jiang, Z.; Wang, J.; Nie, Y. Structural Design and Photoelectron Emission Characteristics of a New Cs2Te Photocathode Based on Surface Plasmon Polaritons. *Radiat. Phys. Chem.* **2024**, *218*, 111545. [CrossRef]
30. Huang, P.-W.; Qian, H.; Du, Y.; Huang, W.; Zhang, Z.; Tang, C. Photoemission and Degradation of Semiconductor Photocathode. *Phys. Rev. Accel. Beams* **2019**, *22*, 123403. [CrossRef]
31. Perrone, A.; Gontad, F.; Lorusso, A.; Di Giulio, M.; Broitman, E.; Ferrario, M. Comparison of the Properties of Pb Thin Films Deposited on Nb Substrate Using Thermal Evaporation and Pulsed Laser Deposition Techniques. *Nucl. Instrum. Methods Phys. Res. A* **2013**, *729*, 451–455. [CrossRef]
32. Mistry, S.; Jones, L.B.; Valizadeh, R.; Hannah, A.; Cropper, M.; Noakes, T.; Militsyn, B.L.; Middleman, K. A comparison of surface properties of metallic thin film photocathodes. In Proceedings of the IPAC, Busan, Republic of Korea, 8–13 May 2016.
33. Liimatainen, J.; Kekkonen, V.; Piirto, J.; Kaisto, J.; Zolotukhin, A.; Chaudhuri, S. Ultra Short Pulsed Laser Deposition Technology for Industrial Applications. *J. Mater. Sci. Eng. B* **2015**, *5*, 196–205.
34. Koskinen, J. Cathodic-Arc and Thermal-Evaporation Deposition. In *Comprehensive Materials Processing*; Hashmi, S., Batalha, G.F., Van Tyne, C.J., Yilbas, B., Eds.; Elsevier: Oxford, UK, 2014.
35. Mohammad, M.A.; Muhammad, M.; Dew, S.K.; Stepanova, M. Fundamentals of Electron Beam Exposure and Development. In *Nanofabrication: Techniques and Principles*; Stepanova, M., Dew, S., Eds.; Springer: Vienna, Austria, 2012; pp. 11–41.
36. Depla, D.; Mahieu, S.; Greene, J.E. Chapter 5—Sputter Deposition Processes. In *Handbook of Deposition Technologies for Films and Coatings*, 3rd ed.; Martin, P.M., Ed.; William Andrew Publishing: Boston, MA, USA, 2010; pp. 253–296.
37. Biasiol, G.; Sorba, L. Molecular Beam Epitaxy: Principles and Applications. In *Crystal Growth of Materials for Energy Production and Energy-Saving Applications*; Edizioni ETS: Pisa, Italy, 2001; pp. 66–83.
38. Chrisey, D.B.; Hubler, G.K. *Pulsed Laser Deposition of Thin Films*; John Wiley & Sons: Hoboken, NJ, USA, 1994.
39. Torrisi, L.; Caridi, F.; Picciotto, A.; Margarone, D.; Borrielli, A. Particle emission from tantalum plasma produced by laser pulse ablation 532 nm. *J. Appl. Phys.* **2006**, *100*, 093306. [CrossRef]
40. Eason, R. *Pulsed Laser Deposition of Thin Films: Applications-Led Growth of Functional Materials*; John Wiley & Sons: Hoboken, NJ, USA, 2007; ISBN 0470052112.
41. Baraldi, G.; Perea, A.; Afonso, C.N. Dynamics of Ions Produced by Laser Ablation of Several Metals at 193 Nm. *J. Appl. Phys.* **2011**, *109*, 43302. [CrossRef]
42. Perea, A.; Gonzalo, J.; Budtz-Joergensen, C.; Epurescu, G.; Siegel, J.; Afonso, C.N.; García-López, J. Quantification of self-sputtering and implantation during pulsed laser deposition of gold. *J. Appl. Phys.* **2008**, *104*, 084912. [CrossRef]
43. Schaber, J.; Xiang, R.; Gaponik, N. Review of Photocathodes for Electron Beam Sources in Particle Accelerators. *J. Mater. Chem. C Mater.* **2023**, *11*, 3162–3179. [CrossRef]
44. Zhao, Q.; Guan, M.Y.; Zhang, P.; Yang, C.G.; Li, J.J.; Wei, Y.T.; Wang, L.; Gan, Y.Y.; Xiong, W.X. Measurement of the Relative Quantum Efficiency of Hamamatsu Model R5912-20MOD Photomultiplier Tubes at Liquid Argon Temperature. *J. Instrum.* **2021**, *16*, T06014. [CrossRef]
45. Bungaro, C.; Noguera, C.; Ballone, P.; Kress, W. Early oxidation stages of Mg (0001): A density functional study. *Phys. Rev. Lett.* **1997**, *79*, 4433. [CrossRef]
46. Mitchell, E.W.J.; Mitchell, J.W. The work functions of copper, silver and aluminium. *Proc. R. Soc. A* **1951**, *210*, 70–84.

47. Wang, X.J.; Rao, T.S.; Batchelor, K.; Ben-Zvi, I.; Fischer, J. Measurements on Photoelectrons from a Magnesium Cathode in a Microwave Electron Gun. *Nucl. Instrum. Methods Phys. Res. A* **1995**, *356*, 159–166. [CrossRef]
48. Gontad, F.; Lorusso, A.; Klini, A.; Manousaki, A.; Perrone, A.; Fotakis, C. Growth of Poly-Crystalline Cu Films on Y Substrates by Picosecond Pulsed Laser Deposition for Photocathode Applications. *Nucl. Instrum. Methods Phys. Res. A* **2015**, *799*, 70–74. [CrossRef]
49. Michaelson, H.B. The work function of the elements and its periodicity. *J. Appl. Phys.* **1977**, *48*, 4729–4733. [CrossRef]
50. Lorusso, A.; Gontad, F.; Broitman, E.; Chiadroni, E.; Perrone, A. Characterisation of Pb Thin Films Prepared by the Nanosecond Pulsed Laser Deposition Technique for Photocathode Application. *Thin Solid Film* **2015**, *579*, 50–56. [CrossRef]
51. Xiang, R.; Schaber, J. Review of Recent Progress on Advanced Photocathodes for Superconducting RF Guns. *Micromachines* **2022**, *13*, 1241. [CrossRef]
52. Iijima, H.; Uesaka, M.; Ueda, T.; Yoshii, K.; Muroya, Y.; Fukasawa, A.; Dobashi, K.; Washio, M.; Kashiwagi, S. High Charge Mg Photocathode RF Gun in S-band Linac at University of Tokyo. In Proceedings of the EPAC'02, Paris, France, 3–7 June 2002; pp. 1771–1773.
53. Nakajyo, T.; Yang, J.; Sakai, F.; Aoki, Y. Quantum Efficiencies of Mg Photocathode under Illumination with 3rd and 4th Harmonics Nd: LiYF4 Laser Light in RF GunJpn. *J. Appl. Phys.* **2003**, *42*, 1470–1774. [CrossRef]
54. Liu, H.-D.; Zhao, Y.-P.; Ramanath, G.; Murarka, S.P.; Wang, G.-C. Thickness Dependent Electrical Resistivity of Ultrathin (<40 Nm) Cu Films. *Thin Solid Film* **2001**, *384*, 151–156.
55. Timoshevskii, V.; Ke, Y.; Guo, H.; Gall, D. The Influence of Surface Roughness on Electrical Conductance of Thin Cu Films: An Ab Initio Study. *J. Appl. Phys.* **2008**, *103*, 113705. [CrossRef]
56. Mistry, S.; Cropper, M.; Valizadeh, R.; Jones, L.B.; Middleman, K.J.; Hannah, A.N.; Militsyn, B.L.; Noakes, T.C.Q. Preparation of polycrystalline and thin film metal photocathodes for normal conducting rf guns. In Proceedings of the IPAC 2015, Richmond, VA, USA, 3–8 May 2015.
57. Lorkiewicz, J.; Nietubyc, R.; Sekutowicz, J.; Barlak, M.; Kostin, D.; Kosinska, A.; Barday, R.; Xiang, R.; Mirowski, R.; Grabowski, W. Review and Present Status of Preparation of Thin Layer Lead Photocathodes for E-Injectors of Superconducting RF Linacs. In Proceedings of the Photonics Applications in Astronomy, Communications, Industry, and High-Energy Physics Experiments, Wilga, Poland, 25–31 May 2015; Volume 9662, pp. 852–861.
58. Sekutowicz, J.; Iversen, J.; Klinke, D.; Kostin, D.; Moller, W.; Muhs, A.; Kneisel, P.; Smedley, J.; Rao, T.; Strzyzewski, P. Status of Nb-Pb Superconducting RF-GUN Cavities. In Proceedings of the 2007 IEEE Particle Accelerator Conference (PAC), Albuquerque, NM, USA, 25–29 June 2007; pp. 962–964.

materials

Article

Nanoindentation Creep Behavior of Additively Manufactured H13 Steel by Utilizing Selective Laser Melting Technology

Evangelos Giarmas [1], Emmanouil K. Tzimtzimis [1], Nikolaos Kladovasilakis [1,2], Dimitrios Tzovaras [2] and Dimitrios Tzetzis [1,*]

[1] Digital Manufacturing and Materials Characterization Laboratory, School of Science and Technology, International Hellenic University, 14th km Thessaloniki-Moudania, Thermi, 57001 Thessaloniki, Greece; e.giarmas@ihu.edu.gr (E.G.)
[2] Centre for Research and Technology Hellas—Information Technologies Institute (CERTH/ITI), 57001 Thessaloniki, Greece
* Correspondence: d.tzetzis@ihu.edu.gr

Abstract: Nowadays, H13 hot work steel is a commonly used hot work die material in the industry; however, its creep behavior for additively manufactured H13 steel parts has not been widely investigated. This research paper examines the impact of volumetric energy density (VED), a critical parameter in additive manufacturing (AM), and the effect of post heat-treatment nitrification on the creep behavior of H13 hot work tool steel, which is constructed through selective laser melting (SLM), which is a powder bed fusion process according to ISO/ASTM 52900:2021. The study utilizes nanoindentation tests to investigate the creep response and the associated parameters such as the steady-state creep strain rate. Measurements and observations taken during the holding phase offer a valuable understanding of the behavior of the studied material. The findings of this study highlight a substantial influence of both VED and nitrification on several factors including hardness, modulus of elasticity, indentation depth, and creep displacement. Interestingly, the creep strain rate appears to be largely unaltered by these parameters. The study concludes with the observation that the creep stress exponent (n) shows a decreasing trend with an increase in VED and the application of nitrification treatment.

Keywords: additive manufacturing; selective laser melting; creep properties; nanoindentation

Citation: Giarmas, E.; Tzimtzimis, E.K.; Kladovasilakis, N.; Tzovaras, D.; Tzetzis, D. Nanoindentation Creep Behavior of Additively Manufactured H13 Steel by Utilizing Selective Laser Melting Technology. *Materials* **2024**, *17*, 3756. https://doi.org/10.3390/ma17153756

Academic Editor: Franz E. Weber

Received: 3 July 2024
Revised: 23 July 2024
Accepted: 26 July 2024
Published: 30 July 2024

1. Introduction

Additive manufacturing (AM) has become a major focus of research in recent years. This process involves building by precisely adding material layer by layer. AM offers advantages over traditional manufacturing, such as faster production, reduced waste, superior customization, and the ability to create near-net shapes. Laser-based AM is particularly useful for metallic parts in various industries. However, AM's inherent complexity, arising from multiple design and process variables, poses challenges for researchers and limits the availability of commercial technologies [1–3].

Driven by the rising demand for AM, particularly selective laser melting (SLM), the need for improved process development, especially for novel powder materials, has intensified. In SLM, a high-power laser melts a layer of powder, building a solid part layer by layer. The laser power, layer thickness, hatch width, and scanning speed are crucial parameters that control the build process and significantly impact both the build rate and the mechanical properties of the final part. Each material requires a specific set of these parameters to achieve desired qualities like high density and surface roughness [4–6]. SLM enables the rapid production of complex, functional metallic parts. Like other AM methods, SLM allows the creation of intricate geometries and features that are challenging or impossible with traditional methods like machining or pressure casting. It also offers the unique ability to manufacture parts with intricate internal cooling channels and lattice structures, and consolidate multiple components into single, functional units [6–13]. This versatility, coupled with the wide availability of various metallic powders, makes SLM a highly attractive AM technology.

To ensure that metal AM parts are suitable for industrial use, it is important to understand their mechanical behavior. This necessitates conducting experiments to assess vital mechanical behavior such as tensile strength and elongation. Numerous studies have already explored the impact of various printing configurations (like process parameters, build orientation, and so on) on the mechanical performance of H13 tool steel components produced through additive manufacturing [14–19]. However, beyond tensile testing, nanoindentation is a commonly employed technique for assessing mechanical properties at submicron levels and holds significant value for experimental studies in the basic physics of materials. It facilitates the identification of individual occurrences like the activation of dislocation sources, the onset of shear instability, and phase changes during the testing process, by leveraging high-precision load-displacement data [20].

Recently, there has been a strong focus on studying the creep behavior and mechanisms of micro- and nanostructured materials and thin films using nanoindentation creep tests [21,22]. Nanoindentation creep experiments offer several distinct benefits over conventional uniaxial creep tests. They can precisely measure localized creep characteristics like those of a thin film applied to a substrate, which cannot be determined using traditional uniaxial creep tests. Additionally, nanoindentation creep methods do not require a significant specimen volume, and multiple repeated tests can be conducted on a single sample. The duration of these tests is also considerably less than that of a standard uniaxial creep test [23]. Nguyen et al. [24] investigated the nanomechanical properties of H13 steel fabricated using the SLM method. The study focused on the interrelationship between the nanoindentation strain rate and hardness. The strain-rate sensitivity exponent (m = 0.022) of this material indicated that the nanoindentation hardness increased in a range of 8.41–9.18 GPa with an increase in the strain rate ranging from 0.002 to 0.1 s^{-1}. In another research, the main focus was the relationship between the nanoindentation strain rate and hardness. The strain-rate sensitivity values were 0.022, 0.019, 0.027, 0.028, and 0.035 for SLM H13 at laser scan speeds of 100, 200, 400, 800, and 1600 mm/s, respectively. This indicates that the hardness increases as the strain rate increases. Notably, the hardness values of the SLM H13 at the 200 mm/s laser scan speed are the highest and least sensitive to the strain rate compared to the H13 samples at other scan speeds [25,26].

In the current paper, the influence of crucial process-related additive manufacturing parameters on the mechanical and creep behavior of H13 hot work tool steel was investigated with the aid of nanoindentation tests. Additionally, the influence of nitrification has been studied. More specifically, the influence of volumetric energy density (VED) on the creep displacement, the creep strain rate, and the creep stress exponent (n) was studied. Other similar studies have utilized a generalized physical model (so-called "Maxwell–Voigt") in order to reveal the plastic origin of high-entropy metallic glasses [27]. The presented research is very critical, because not many similar studies are available in the literature concerning AM H13 hot work tool steel. The primary objective of this study is the investigation of the creep behavior of additively manufactured H13 hot work tool steel, with the intent of integrating it into standard industrial operations as a substitute for traditionally produced steel. Given the lack of comparable research, this investigation is poised to yield numerous significant findings. Figure 1 depicts the applied methodology of the present research paper.

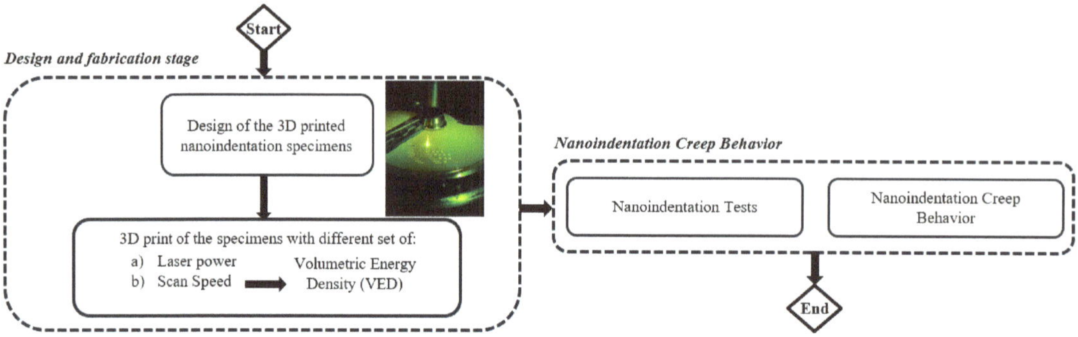

Figure 1. Flowchart of the current research work.

2. Material and Methods

2.1. Materials and Sample Preparation

The material that has been selected to be investigated is the OERLIKON MetcoAdd™ H13-A (Westbury, NY, USA), an air-hardenable, iron-chromium martensitic steel powder. Its chemical composition according to the manufacturer is shown in Table 1.

Table 1. PSD, Hall flow, and chemical composition for H13A powder according to the manufacturer.

Nominal Range (μm)		D90 (μm)	D50 (μm)	D10 (μm)	Hall Flow (s/50 g)
−45 + 15		50	34	21	≤25
Weight Percent (nominal)					
Fe	Cr	Mo	Si	V	C
Balance	5.2	1.3	1.0	1.0	0.4

To achieve the desired mechanical behavior of H13 hot work tool steel, various printing parameters were considered. The energy imparted to the material is a very important variable in the SLM process, because it directly affects the characteristics of the molten pool during formation and, consequently, the properties of the final part. This energy is known as volumetric energy density (VED), measured in J/mm^3, and is described as follows:

$$\mathrm{VED} = \frac{\mathrm{P}}{\mathrm{V} * \mathrm{h} * \mathrm{t}}\left[\frac{\mathrm{J}}{\mathrm{mm}^3}\right] \tag{1}$$

where P is the laser power (J/s), V is the scanning speed (mm/s), h is the hatch width (mm), and t is the layer thickness (mm). Katancik et al. explored a range of printing parameters [28]. The component with the highest density, exhibiting a relevant density of 99% and a porosity of less than 0.01%, was fabricated utilizing a volumetric energy density of 760 J/mm^3. This was accomplished with settings of 152 W laser power, 100 mm/s scanning speed, 40 mm hatch spacing, and 50 mm layer thickness. In the current work, the hatch width was selected constant at 40 μm, and the layer thickness at 25 μm. The variables were the scanning speed and the laser power. The different combinations of printing parameters are presented in Table 2.

Table 2. Printing parameters for each set of testing specimens.

Sample	VED (J/mm^3)	Laser Power (W)	Scanning Speed (mm/s)
S1	50	100	2000
S2	57	100	1750
S3	58	115	2000
S4	66	115	1750
S5	68	135	2000
S6	75	150	2000
S7	77	135	1750
S8	80	100	1250
S9	86	150	1750
S10	92	115	1250
S11	100	100	1000
S12	108	135	1250
S13	115	115	1000
S14	120	150	1250
S15	135	135	1000
S16	150	150	1000

2.2. Nanoindentation Test

Nanoindentation experiments were conducted using a Shimadzu DUH-211S instrument (Kyoto, Japan) equipped with a Berkovich diamond tip (included angle: 65°, tip radius: 100 nm). Cylindrical specimens were used for the conducted nanoindentation tests. The calculation method of Oliver and Pharr has been used in order to determine the indentation hardness and the elastic modulus of the SLM printed parts [29]. The indentation's maximum depth in the following function can be used in order to determine the hardness (H):

$$H = \frac{P_{max}}{A} \tag{2}$$

where P_{max} is the maximum applied load measured at the maximum depth of penetration (h_{max}), and A is the projected contact area between the indenter and the film. For a perfect Berkovich indenter, A can be calculated as a function of the contact indentation depth h_f as

$$A = 33\sqrt{3}h_f^2 \tan^2 65 = 23.96h_f^2 \tag{3}$$

The contact indentation, known as h_f, can be calculated using the following formula:

$$h_f = h_{max} - \varepsilon\frac{P_{max}}{S} \tag{4}$$

where ε is a geometric constant that takes a value of 0.75 for an indenter with the shape of a pyramid, and S is the contact stiffness. The contact stiffness, S, can be revealed as the gradient of the unloading curve at the point of maximum load as

$$S = \left(\frac{dP}{dh}\right)_{h=h_{max}} \tag{5}$$

The reduced elastic modulus E_r is given by

$$E_r = \frac{S}{2\beta}\sqrt{\frac{\pi}{A}} \tag{6}$$

where β is a constant that is determined by the shape of the indenter. For the Berkovich indenter used in this case, the value of β is set to 1.034. Following this, the elastic modulus of the specimen (E_S) can be computed as per the following formula:

$$\frac{1}{E_r} = \frac{1-v_s^2}{E_s} + \frac{1-v_i^2}{E_i} \tag{7}$$

where $v_{i,s}$ and $E_{i,s}$ describe the Poisson's ratio and elastic modulus, respectively. Furthermore, for a diamond indenter, the Poisson's ratio (v_i) is 0.07 and the elastic modulus (E_i) is 1140 GPa. The specimen's elastic modulus (E_S) and hardness (H) are calculated using the equations mentioned above.

The instrument possesses a load resolution of 0.196 μN. All tests were performed at room temperature. The nanoindentation procedure involved applying a controlled load profile to the surface of the filaments via the diamond tip. The load profile consisted of a continuous increase to a peak value of 800 mN, held constant for a dwell time of 3 s (creep time), followed by a complete unload to zero force. The indentation depth was continuously monitored as a function of the applied load. The average of 10 individual measurements was employed to determine the elastic modulus and hardness of the material.

2.3. Nitrification

Concerning the next step of the nanoindentation experiments, the specimens underwent surface treatment via nitridation to investigate their influence on mechanical behavior. The nitrification program employed a two-stage cooling process and it is depicted in Figure 2. The program encompassed a total duration of 20 h and 11 min. During the initial phase (4 h and 12 min), the temperature steadily increased to reach the target value of 397 °C. Subsequently, ammonia gas was introduced into the furnace chamber (4 h and 42 min). The temperature was then elevated further to 500 °C within the next hour (5 h and 46 min). This stage marked the initiation of the first nitriding step, which concluded at 8 h and 25 min. The second stage commenced with another temperature ramp, reaching a peak of 530 °C at 13 h and 24 min. This stage concluded the nitriding process. A two-step cooling sequence followed. The first stage reduced the temperature to 255 °C by 15 h and 24 min. The final cooling stage brought the temperature down to 70 °C, signifying the completion of the nitridation program at 20 h and 11 min.

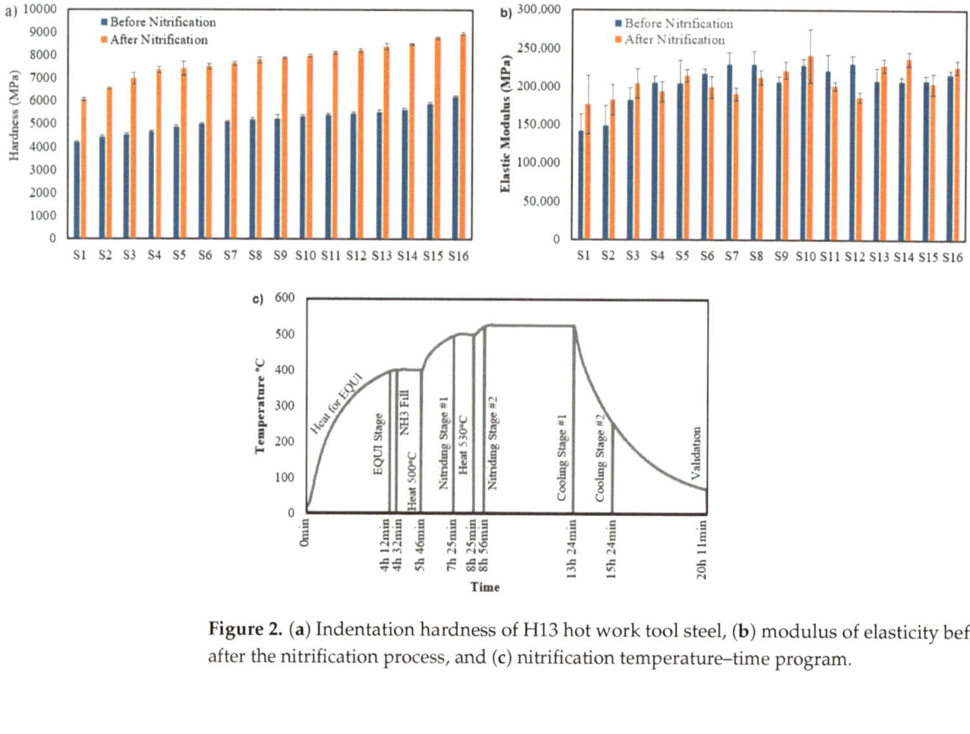

Figure 2. (**a**) Indentation hardness of H13 hot work tool steel, (**b**) modulus of elasticity before and after the nitrification process, and (**c**) nitrification temperature–time program.

3. Results and Discussion

3.1. Nanoindentation Test Results

Figure 2 depicts the results for the hardness and the modulus of elasticity. For all the selected parameters, 10 sets of specimens were fabricated for each printing parameter and the standard deviation of them is also depicted in this figure. Specimen 16 demonstrated the utmost hardness, reaching 6191 ± 60 MPa. It becomes clear that as the VED increases, there is a corresponding rise in hardness. The greatest modulus of elasticity was observed in specimens 6 through 14 and 16, each surpassing 215,000 MPa. On the other hand, specimens 1 and 2 presented the minimal elastic modulus, falling below 150,000 MPa. These reduced modulus values are due to the use of varied printing parameters, which were inadequate for achieving the required mechanical performance standards. In terms of indentation depth, specimen 16 showed the most advantageous properties, with an indentation depth of 2.11 ± 0.01 µm. The largest indentation depth was recorded for specimen 1 at 2.7 ± 0.01 µm, a result of choosing additive manufacturing parameters that did not meet the necessary mechanical performance benchmarks.

Nitrification significantly enhanced the nanoindentation behavior of H13 hot work tool steel, making it more resistant to the extreme conditions of extrusion. Hardness increased by up to 51.23% (Table 3), with specimen 16 exhibiting the highest value (8960 ± 77 MPa) after treatment. The effect on elastic modulus was less pronounced, with an average increase of around 4.3%. However, all specimens displayed a modulus exceeding 215,000 MPa.

Table 3. Comparative nanoindentation results for additively manufactured H13 hot work steel before and after the nitrification process.

VED	Hardness (MPa)		Difference	Elastic Modulus (MPa)		Difference
	Before	After		Before	After	
50	4196.78	6076.76	44.80%	141,520	177,000	25.07%
57	4444.36	6555.67	47.51%	148,280	182,780	23.27%
58	4553.83	6989.61	53.49%	182,900	204,500	11.81%
66	4666.16	7371.27	57.97%	205,100	193,300	−5.75%
68	4880.79	7431.27	52.26%	204,260	214,740	5.13%
75	4992.83	7515.16	50.52%	217,460	199,380	−8.31%
77	5093.64	7658.96	50.36%	229,140	190,700	−16.78%
80	5189.16	7810.58	50.52%	228,860	212,260	−7.25%
86	5235.03	7896.8	50.85%	205,900	221,740	7.69%
92	5334.36	8001.45	50.00%	227,780	240,820	5.72%
100	5386.71	8137.8	51.07%	221,200	201,540	−8.89%
108	5459.57	8232.46	50.79%	230,000	186,680	−18.83%
115	5548.13	8390.42	51.23%	208,280	228,700	9.80%
120	5648.06	8509.95	50.67%	207,360	236,780	14.19%
135	5903.83	8789.17	48.87%	208,440	204,220	−2.02%
150	6191.25	8960.81	44.73%	216,360	226,660	4.76%

3.2. Microstructural Characterization

The microstructure of the additively manufactured metal parts was investigated using a Phenom ProX instrument (ThermoFisher Scientific, Waltham, MA, USA) scanning electron microscope (SEM) equipped with both a backscattered electron detector (BSD) and a secondary electron detector (SED) for enhanced characterization. This combination of detectors ensured a comprehensive analysis of the microstructure. Figure 3a illustrates the microstructure of specimens fabricated with a low energy density. Large areas of unmelted powder particles are evident, indicating incomplete fusion with the surrounding material. These unfused regions contribute to surface irregularities and microstructural discontinuities, which could potentially have a detrimental effect on the mechanical performance of the final parts. To illustrate this point further, Figure 3a presents higher magnification SEM images of specimens produced with a low energy density (50 J/mm^3). Unmelted particles are clearly visible within the microstructure. Conversely, as the energy density during printing increased, the presence of these unfused regions diminished. Figure 3b,c, corresponding to energy densities of 100 J/mm^3 and 150 J/mm^3, respectively, demonstrate a significant reduction in unmelted particles. This improved melting efficiency translates to smoother surfaces and potentially enhanced mechanical properties.

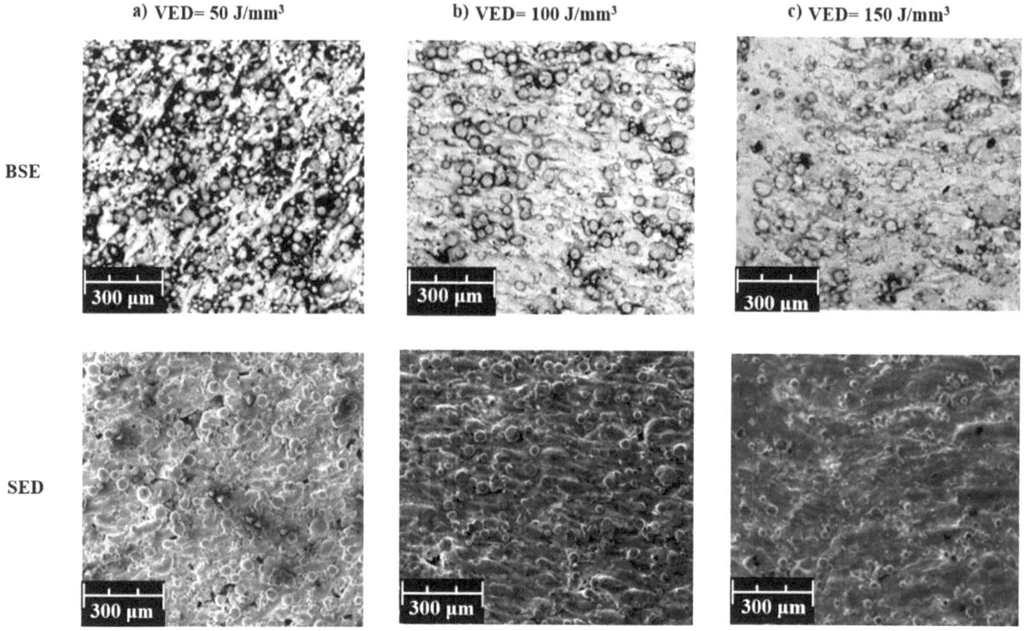

Figure 3. SEM images of the specimens: (**a**) low energy density (50 J/mm^3), (**b**) medium energy density (100 J/mm^3), (**c**) high energy density (150 J/mm^3).

3.3. Nanoindentation Creep Behavior

In order to study the creep behavior with nanoindentation tests, the data collected during the holding stage of the experiment were used. The experimental data during the holding stage could be well-fitted with the following equation [16]:

$$h = hi + a(t\text{-}ti)1/2 + b(t\text{-}ti)1/4 + c(t\text{-}ti)1/8 \quad (8)$$

where h is the indenter displacement during the holding stage; t is creep time; and hi, ti, a, b, and c are the best-fit parameters that came from Equation (8).

The experimental data and fitted creep displacement and time curve for S1 with VED of 50 J/mm^3 without nitrification during the nanoindentation holding stage are typically depicted in Figure 4. The red line represents the fitted curve according to Equation (8) and the blue dots show the experimental results. Furthermore, Table 4 shows the values of the best-fit parameters from Equation (8). It is clearly revealed that the fitted curve presents an almost perfect agreement with the experimental results for the ambient-temperature creep of the H13 additively manufactured hot work tool steel during the holding stage of the nanoindentation test. The best-fit parameters have been found with the aid of a command line–driven graphing utility (Gnuplot/www.gnuplot.info, accessed from 2 September 2022 until 15 December 2022), where the parameters that satisfy Equation (8) according to the collected data from the experiments have been calculated.

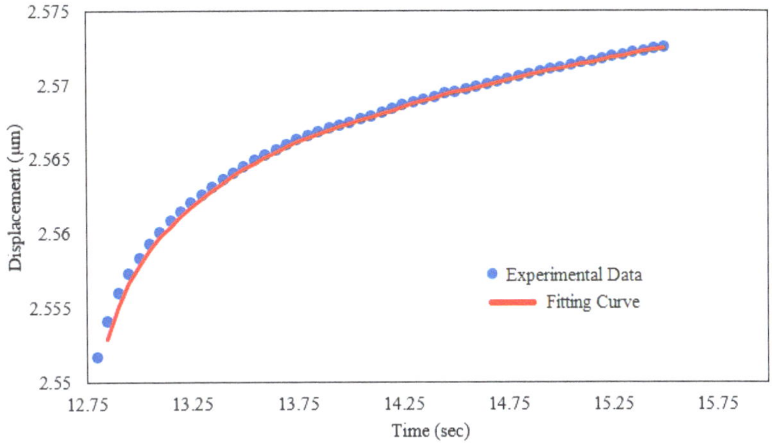

Figure 4. Experimental and fitted creep displacement and time curve for Sample 1 with a VED of 50 J/mm^3 without nitrification.

Table 4. Values of the best-fit parameters and asymptotic Standard Error S1 with a VED of 50 J/mm^3 without nitrification.

Set of Parameters	Asymptotic Standard Error	
a = 0.00538866	±0.001072	(19.89%)
b = 0.019735	±0.003034	(15.38%)
c = −0.0144336	±0.002013	(13.95%)
hi = 2.55421	±0.0002216	(0.0087%)
ti = 12.6707	±0.01187	(0.09366%)

The research data presented in Figure 5 illustrate the creep displacement–time curves, which were derived through the process of nanoindentation. These curves represent the outcomes for 16 distinct conditions of volumetric energy density (VED), both with and without the application of a nitrification heat-treating process, as detailed in Table 5. The results of this study underscore the substantial impact of both VED and nitrification on the creep behavior of the material under investigation. In the case of specimens that did not undergo nitriding, an increase in VED from 50 J/mm^3 to 150 J/mm^3 led to a notable decrease of 31.32% in creep displacement, changing from 0.0166 μm to 0.0114 μm. Similarly, specimens that were subjected to nitriding displayed a 33.79% reduction in creep displacement, decreasing from 0.0145 μm to 0.0096 μm, when the same VED values were applied. It is worth noting that the process of nitrification consistently resulted in a reduction of creep displacement across all VED conditions, with an average decrease observed to be 24.54%.

These findings highlight the synergistic effect of VED and nitrification on enhancing the creep resistance of H13 hot work tool steel, which was fabricated using the selective laser melting (SLM) technique. The combination of a VED of 150 J/mm^3 and the application of a nitrification treatment resulted in the lowest observed creep displacement, indicating the potential for optimizing the material's performance in this regard.

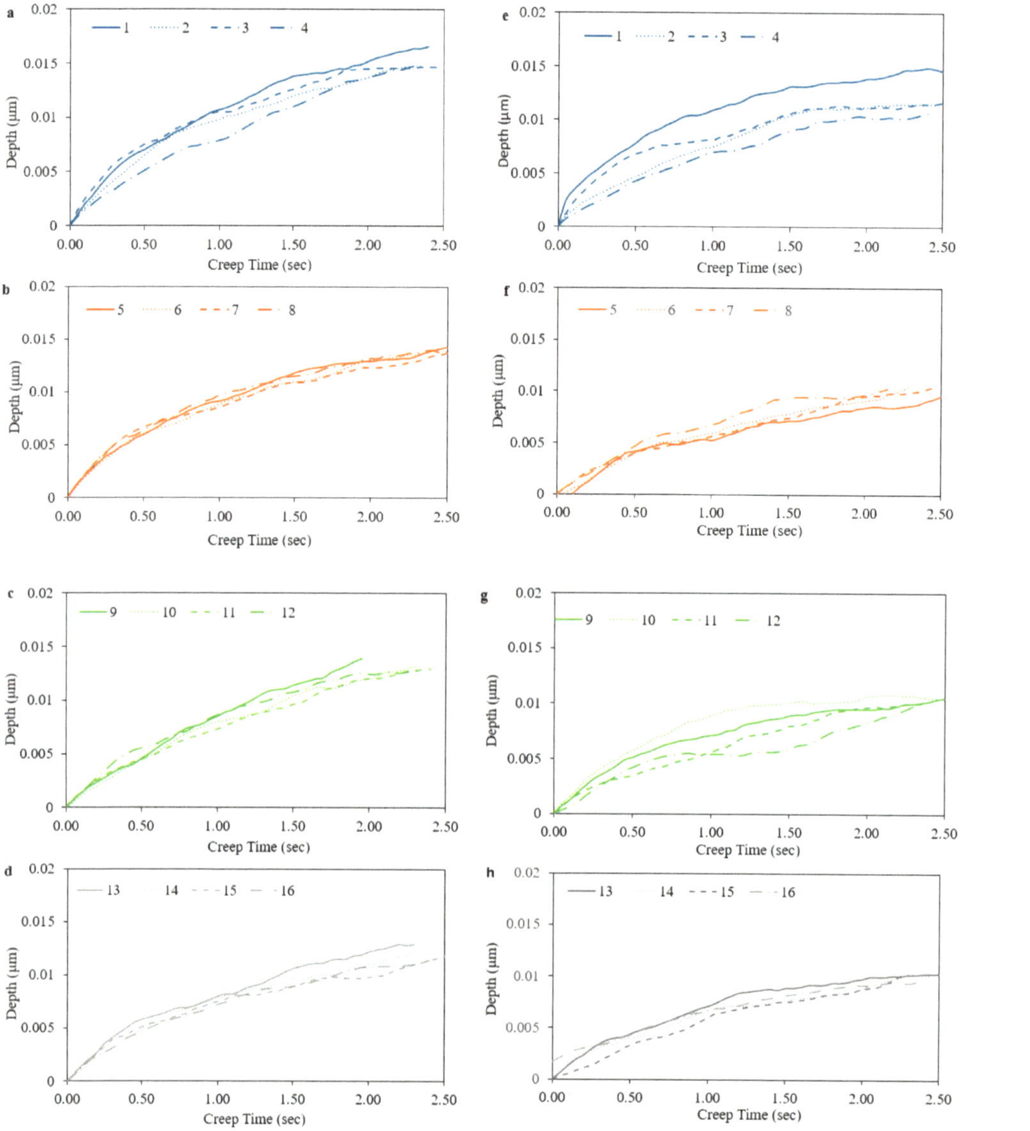

Figure 5. Creep displacement–time curves for different VED conditions without (**a–d**)/with (**e–h**) nitrification.

According to Mayo et al. [19], the creep strain rate $\dot{\varepsilon}$ was calculated from displacement through the following equation:

$$\dot{\varepsilon} = \frac{\dot{h}}{h} = \frac{1}{h}\frac{dh}{dt} \tag{9}$$

The creep strain rate data were calculated by differentiating the fitted creep h-t curves in the hold stage with Equation (8) while Figures 6 and 7 depict the creep strain-time curves for the 16 selected volumetric energy density (VED) conditions (S1–S16), both with and without the nitrification heat-treating process. It can be seen that the creep $\dot{\varepsilon}$-t curves consist of the following two stages: a transient one and a steady-state one. At the beginning of the load holding, a very high creep strain rate was measured and then the creep strain rate decreased rapidly. This phenomenon may occur due to work hardening caused by the plastic deformation, corresponding to the transient creep. After a holding time of 2.3 s, an almost steady-state creep phase is depicted and a very slow decrease in the creep strain rate with increasing creep time is revealed [20]. From Figures 6 and 7, it is evident that there is no important change concerning the creep strain rate over the creep time for both VED and nitrification, but a small decrease in the creep strain rate with the use of nitrifications can be detected. Liu et al. concluded to the same insensitivity concerning the steady creep strain rate due to the dislocation creep phenomenon [30]. In each diagram, a detailed image has been superimposed, where the difference between the different aging conditions is more clearly visible.

Table 5. Creep displacement for different VED conditions.

Sample Number	VED (J/mm^3)	Creep Displacement (μm) *	% Change	Creep Displacement (μm) **	% Change
S1	50	0.0166	-	0.0145	-
S2	57	0.0148	−10.84%	0.0119	−17.93%
S3	58	0.0149	0.68%	0.0117	−1.68%
S4	66	0.0148	−0.67%	0.0108	−7.69%
S5	68	0.0143	−3.38%	0.0103	−4.63%
S6	75	0.0142	−0.70%	0.0105	1.94%
S7	77	0.014	−1.41%	0.0104	−0.95%
S8	80	0.0141	0.71%	0.0102	−1.92%
S9	86	0.0139	−1.42%	0.0106	3.92%
S10	92	0.0132	−5.04%	0.0104	−1.89%
S11	100	0.0129	−2.27%	0.0101	−2.88%
S12	108	0.013	0.78%	0.01	−0.99%
S13	115	0.0129	−0.77%	0.0103	3.00%
S14	120	0.0121	−6.20%	0.0101	−1.94%
S15	135	0.0119	−1.65%	0.0102	0.99%
S16	150	0.0114	−4.20%	0.0096	−5.88%

* without nitrification; ** with nitrification.

The dominant creep mechanism and the ambient-temperature creep behavior of the materials can be studied by Equation (10) of the stress (s) and the creep strain rate ($\dot{\varepsilon}$) [21]:

$$\dot{\varepsilon} = \alpha\sigma^n \quad (10)$$

where n is the creep stress exponent and α is a material constant. The stress in a nanoindentation test should be connected to the pressure applied by the indenter. From the relationship of $H = P/A = P/24.56h_c^2$ (where P is the holding load and A is the projected contact area of the Berkovich tip), the following equation for the creep stress in the holding stage can be easily assumed based on $\sigma = (H/3)/(h_{max}/h)^2$ [22]. Moreover, the creep stress exponent n is derived by determining the slope of the ln $\dot{\varepsilon}$ versus lnσ plot according to Equation (11):

$$n = \frac{d(ln\dot{\varepsilon})}{d(ln\sigma)} \quad (11)$$

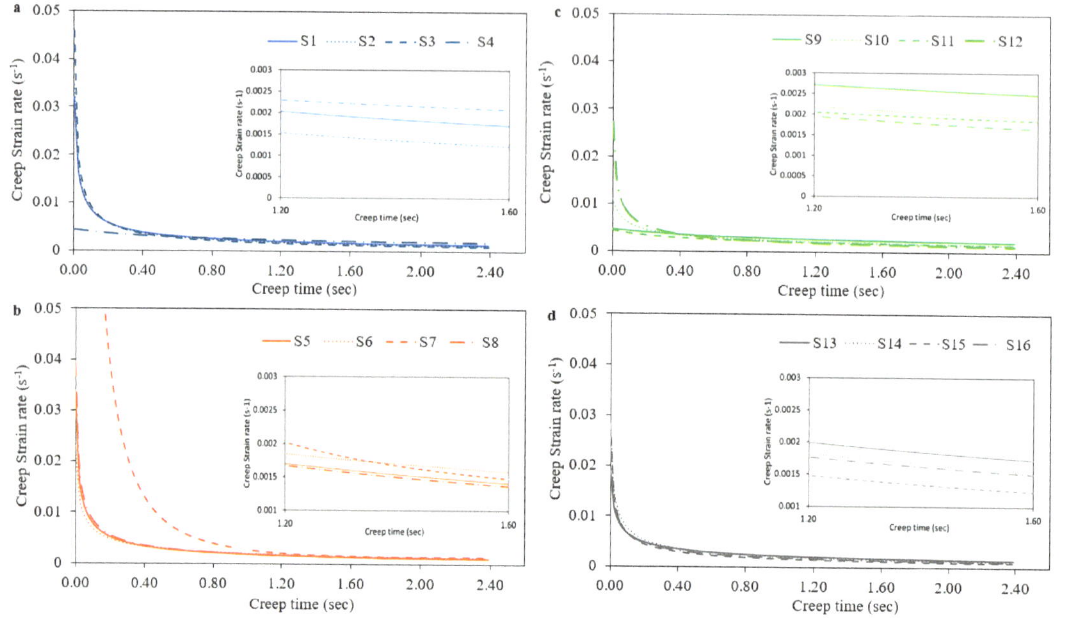

Figure 6. Creep strain–time curves for different VED conditions without nitrification (**a**) S1–S4, (**b**) S5–S8, (**c**) S9–S12 and (**d**) S13–S16.

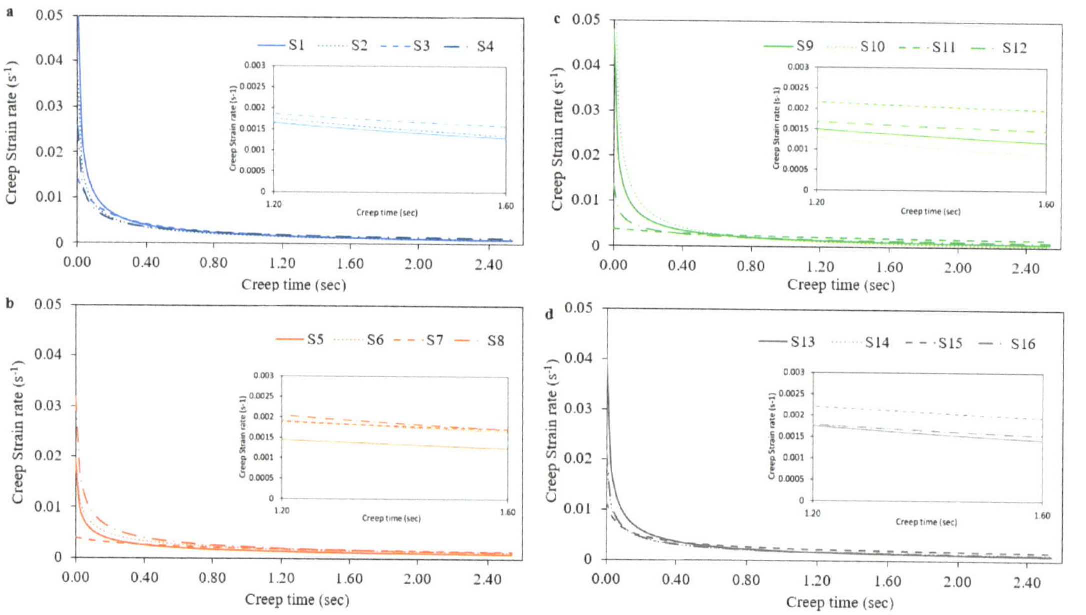

Figure 7. Creep strain–time curves for different VED conditions with nitrification (**a**) S1–S4, (**b**) S5–S8, (**c**) S9–S12 and (**d**) S13–S16.

Figure 8 depicts the $ln\dot{\varepsilon}$ versus $ln\sigma$ diagrams, using the definitions of creep strain rate and stress, for the 16 selected volumetric energy density (VED) conditions, both with and without the nitrification heat-treating process. In order to make the calculations for each one of the selected specimens according to the different VED conditions and the use of nitrification as a heat-treating process, only the steady-state creep is taken into consideration because the value of the creep stress exponent differs at different locations in the creep strain rate vs. stress curve.

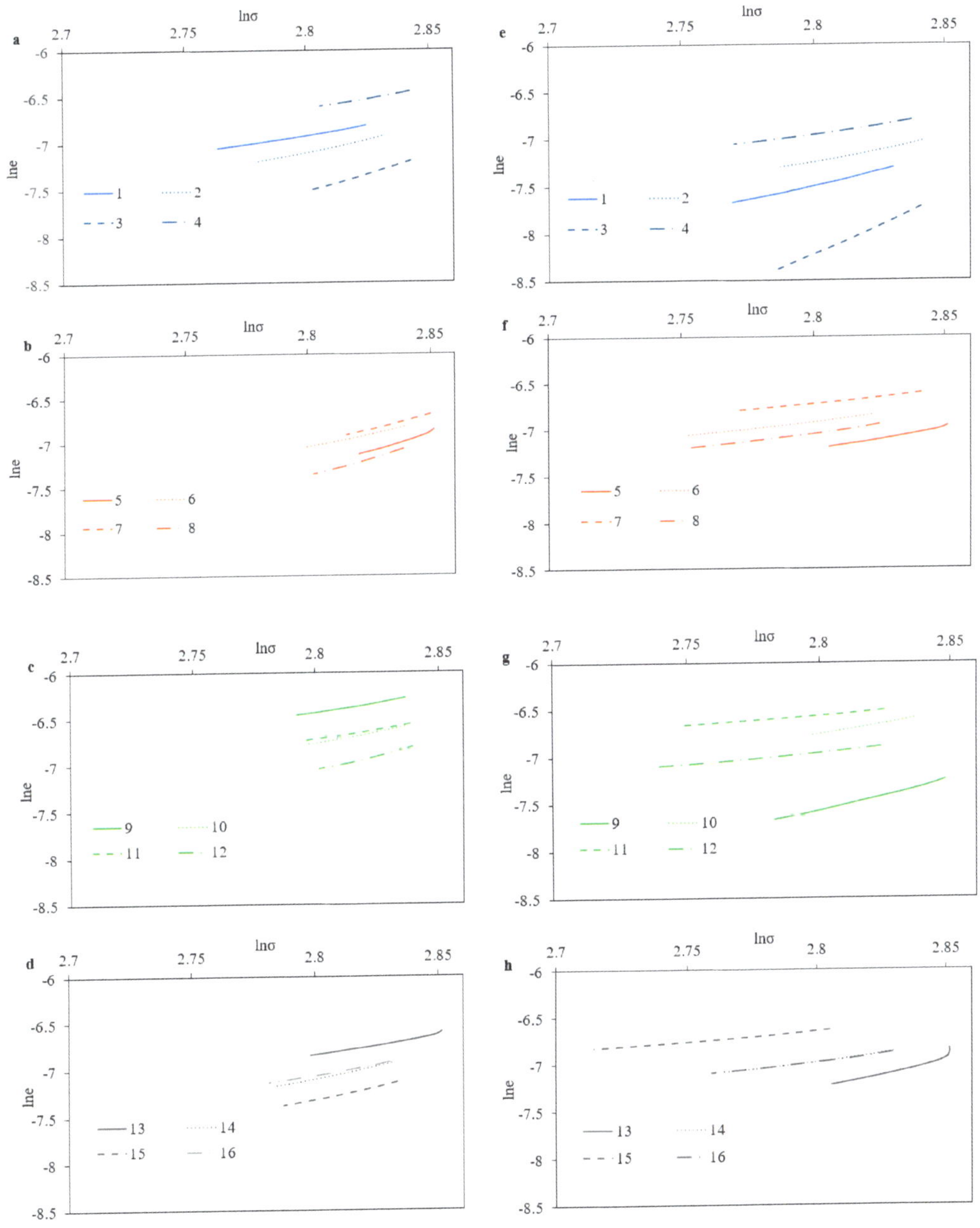

Figure 8. The ln$\dot{\varepsilon}$-lnσ curves for different VED conditions without (**a**–**d**)/with (**e**–**h**) nitrification.

Furthermore, the values of the steady creep strain rate ($\dot{\varepsilon}$) and the creep stress exponent (n) for the different VED conditions without and with the stage of nitrification are shown in Tables 6 and 7, respectively. It is evident that the steady creep strain rate does not vary significantly with the VED values. The values are very low and as a result very sensitive to change that it will not be safe to make a clear assumption. The main reason for this may be that despite the values of VED, there might be important differences to the printing parameters from which the VED are revealed. This might be a good starting point for future research. However, concerning the creep stress exponent (n), it seems that with increasing the VED from 50 J/mm^3 to 150 J/mm^3, there is a total decrease of 43.32% (from 6.67 to 3.78) for non-nitrided specimens. Similarly, the nitrided specimens exhibited a 50.73% reduction (from 5.48 to 2.70). Finally, nitrification consistently reduced the stress exponent (n) across the majority of VED conditions, with an average decrease of 2.94%.

Table 6. Steady creep strain rate ($\dot{\varepsilon}$) for different VED conditions.

Sample Number	VED (J/mm^3)	Steady Creep Strain Rate (e) *	% Change	Steady Creep Strain Rate (e) **	% Change
S1	50	0.00088045	-	0.000481014	-
S2	57	0.000745593	−15.32%	0.000667245	38.72%
S3	58	0.000542276	−27.27%	0.000226144	−66.11%
S4	66	0.001308202	141.24%	0.000833911	268.75%
S5	68	0.000716387	−45.24%	0.000682166	−18.20%
S6	75	0.000872306	21.76%	0.000856347	25.53%
S7	77	0.000953821	9.34%	0.001076213	25.67%
S8	80	0.000655186	−31.31%	0.000763662	−29.04%
S9	86	0.001507832	130.14%	0.000433518	−43.23%
S10	92	0.001126749	−25.27%	0.001126749	159.91%
S11	100	0.001182228	4.92%	0.001257004	11.56%
S12	108	0.000862596	−27.04%	0.000811422	−35.45%
S13	115	0.000953821	10.58%	0.0006205	−23.53%
S14	120	0.000760173	−20.30%	0.000833777	34.37%
S15	135	0.000634656	−16.51%	0.001090283	30.76%
S16	150	0.000799513	25.98%	0.000830285	−23.85%

* without nitrification; ** with nitrification.

Table 7. Creep stress exponent (n) for different VED conditions.

VED (J/mm^3)	Creep Stress Exponent (n) *	% Change	Creep Stress Exponent (n) **	% Change
50	6.675110931	-	5.486328588	-
57	4.199017749	−37.09%	4.501311944	−17.95%
58	3.229997014	−23.08%	11.39958664	153.25%
66	3.485868021	7.92%	2.971131928	−73.94%
68	5.86734502	68.32%	3.290291092	10.74%
75	4.538255719	−22.65%	2.687519724	−18.32%
77	4.710518564	3.80%	2.031330012	−24.42%
80	6.566776868	39.41%	3.036948594	49.51%
86	2.950171409	−55.07%	5.068478173	66.89%
92	3.842315907	30.24%	4.180542552	−17.52%
100	3.274040411	−14.79%	1.773010121	−57.59%
108	4.855511694	48.30%	1.986774532	12.06%
115	2.568365618	−47.10%	4.159389552	109.35%
120	4.111052477	60.06%	2.877173982	−30.83%
135	4.44471448	8.12%	1.850538219	−35.68%
150	3.780465456	−14.94%	2.704244278	46.13%

* without nitrification; ** with nitrification.

4. Conclusions

The effect of volumetric energy density (VED) and the use of nitrification as a heat-treating process on the creep behavior of H13 hot work tool steel produced with the SLM process, and relevant creep parameters such as steady creep strain rate ($\dot{\varepsilon}$) and the creep stress exponent (n), were investigated in the presented research using nanoindentation tests. Concerning the hardness, the modulus of elasticity, and the indentation depth, specimen 16 with a VED of 150 J/mm^3 exhibits the most favorable characteristics both before and after the nitrification process. Nitrification significantly enhanced the nanoindentation behavior of H13 hot work tool steel. Hardness increased by up to 51.23%, with specimen 16 exhibiting the highest value (8960 ± 77 MPa) after heat treatment. The effect on the elastic modulus was less pronounced, with an average increase of around 4.3%. However, all specimens displayed a modulus exceeding 215,000 MPa after nitrification. Moreover, a significant influence of both VED and nitrification on creep behavior has been revealed. Non-nitride specimens displayed a 31.32% reduction in creep displacement (from 0.0166 μm to 0.0114 μm) as the VED increased from 50 J/mm^3 to 150 J/mm^3. Similarly, nitride specimens exhibited a decrease in creep displacement of 33.79% (from 0.0145 μm to 0.0096 μm) at these VED levels. Furthermore, nitrification consistently reduced creep displacement across all VED conditions, with an average decrease of 24.54%. Finally, despite that the steady creep strain rate does not vary significantly with the aging process, the creep stress exponent (n) seems to decrease by 43.32% (from 6.67 to 3.78) for non-nitrided specimens and 50.73% for the nitride ones.

To sum up, the aforementioned studies pave the path for initiating considerations of the creep behavior in additively manufactured H13 steel components. Given the criticality of creep resistance in the high-performance applications typically associated with these metals, understanding this behavior is paramount.

Author Contributions: Conceptualization, E.G. and D.T. (Dimitrios Tzetzis); methodology, E.G. and D.T. (Dimitrios Tzetzis); software, E.G.; validation, N.K. and E.K.T.; formal analysis, E.G. and E.K.T.; investigation, E.G.; resources, E.K.T., N.K. and D.T. (Dimitrios Tzovaras); data curation, E.G. and E.K.T.; writing—original draft preparation, E.G.; writing—review and editing, E.G., N.K., D.T. (Dimitrios Tzovaras) and D.T. (Dimitrios Tzetzis).; visualization, E.G. and N.K.; supervision, D.T. (Dimitrios Tzetzis); project administration, D.T. (Dimitrios Tzetzis). All authors have read and agreed to the published version of the manuscript.

Funding: This research received no external funding.

Institutional Review Board Statement: Not applicable.

Informed Consent Statement: Not applicable.

Data Availability Statement: The raw data supporting the conclusions of this article will be made available by the authors on request.

Conflicts of Interest: The authors declare no conflicts of interest.

References

1. Payne, G.; Ahmad, A.; Fitzpatrick, S.; Xirouchakis, P.; Ion, W.; Wilson, M. Remanufacturing H13 Steel Moulds and Dies Using Laser Metal Deposition. In Proceedings of the 14th International Conference on Manufacturing Research, Loughborough, UK, 6–8 September 2016. [CrossRef]
2. Gibson, I.; Rosen, D.W.; Stucker, B. *Additive Manufacturing Technologies: 3D Printing, Rapid Prototyping, and Direct Digital Manufacturing*, 2nd ed.; Springer: Berlin/Heidelberg, Germany, 2015.
3. Sames, W.J.; List, F.; Pannala, S.; Dehoff, R.R.; Babu, S.S. The metallurgy and processing science of metal additive manufacturing. *Int. Mater. Rev.* **2016**, *61*, 1–46. [CrossRef]
4. Kladovasilakis, N.; Charalampous, P.; Tsongas, K.; Kostavelis, I.; Tzovaras, D.; Tzetzis, D. Influence of Selective Laser Melting Additive Manufacturing Parameters in Inconel 718 Superalloy. *Materials* **2022**, *15*, 1362. [CrossRef] [PubMed]
5. Kladovasilakis, N.; Charalampous, P.; Kostavelis, I.; Tzovaras, D.; Tzetzis, D. Impact of metal additive manufacturing parameters on the powder bed fusion and direct energy deposition processes: A comprehensive review. *Prog. Addit. Manuf.* **2021**, *6*, 349–365. [CrossRef]

Nevertheless, Inconel is a hard-to-machine material [4], and the heat generated by machining is excessive due to its high toughness and strength, even at high temperatures. Due to its low thermal conductivity, heat is concentrated on the tool surface during the machining of Inconel, which leads to increased tool wear. During the machining of these alloys, the temperature and stresses near the cutting tool edge are significantly high, making it susceptible to damage such as flank wear, notching wear, and edge chipping. These issues result in a reduced tool lifespan and a decline in the integrity of the finished surface [5]. Additionally, welding and adhesion of the machined material to the cutting tool frequently occur, causing severe notching and altering the tool's rake face due to the subsequent pull-out of the coating and tool substrate [6]. Therefore, various cooling and lubrication methods are used in the milling process to reduce cutting forces and tool wear. For nickel alloys, a great variety of methods such as dry cutting, flood cutting, cryogenic cutting, and cryogenic minimum quantity lubrication can be used for this purpose [7]. Despite the variety of methods, cutting flood cutting using water-oil emulsions with a concentration of about 3–9% or minimal quantity lubrication (MQL) is most common in industry [8]. Based on this, it was concluded that the use of cutting fluids in both the combined DMD and milling process will be unavoidable for hard-to-machine materials in most cases. It may lead to interface contamination in the sequential surfacing and milling process. The following sections present research on the combined process, highlighting the significance of ongoing investigations in this field and reviewing the current state of the art.

Researchers [9–13] studied the machinability of IN718 produced by AM methods. Their findings showed the influence of the porosity and heat treatment on the cutting forces, tool wear, or surface roughness. These results indicate a strong interest in the topic of machining parts produced by AM methods.

Song et al. [14] reported on a hybrid approach called "3D welding and milling" that combines gas metal arc welding with milling. The study focuses on the early stages of process development and characterises parts fabricated using this approach. It delves into the results, highlighting the unique capabilities of this hybrid process in fabricating metallic prototypes and potentially overcoming limitations of surface finish and accuracy often associated with additively manufactured parts.

Soffel et al. [15] focused on the fabrication of parts using casting, interface milling, and DMD. They reveal that different processing routes impact the mechanical properties and bonding quality of the final product. This highlights the importance of understanding how DMD and milling interact at the interface level.

Authors like Barckhoff [16] and Alsher et al. [17] have noted that before laser welding, any lubricants on the surface should be cleaned off to prevent the formation of gas bubbles and pores. Cortina et al. [18] analysed the impact of using cutting fluid in a hybrid manufacturing process that combines additive and machining operations of IN718. The paper examines whether DMD after machining is feasible without requiring intermediate cleaning to remove residual cutting fluids. It concludes that if the part surface is not adequately cleaned after machining, it could lead to poor quality in the subsequent laser additive operation. However, in this work, only the porosity of the prepared samples was evaluated. In addition, the samples themselves are only single-layered on a forged substrate, which does not fully reflect the real combined manufacturing process of 3D parts, which is a multilayer cladding on an already deposited part of the workpiece.

It is known that cutting fluids, remaining in the zone of interaction with the laser during further cladding, are subjected to heating, vaporisation, and dissociation, which can accelerate the diffusion of hydrogen or carbon into the material, leading to undesirable effects such as hydrogen embrittlement. Aiello et al. [19] investigated the effect of H content in IN718 manufactured by Selective Laser Melting (SLM). Tests were carried out at hydrogen concentrations up to 28 parts per million (ppm). The results show hydrogen embrittlement at hydrogen concentrations starting from 2.7 ppm (drop in elongation at break by about 10%). This suggests a significant effect of hydrogen concentration on the properties of IN718.

Thus, the problem of using coolant in hybrid processes is an important issue that requires further study. Research in this direction can assist in determining the need for an additional technological cleaning operation, which is likely to increase the cost of the process and the final product. The aim of this study is to investigate the effect of interface cleaning methods in a combined DMD + milling process on the quality of the interface itself by studying the porosity, carbon and hydrogen content on multi-layer samples. To the best of the authors' knowledge, the impact of liquids on the subsequential DMD process has not been investigated in such detail previously.

2. Materials and Methods

2.1. Experimental Setup

The nickel-based superalloy Inconel 718 (Oerlikon Metco AG, Oerlikon, Switzerland) was selected for the experiments. A 5-axis combined milling and DMD machine GF HPM 450 U (Georg Fischer AG, Schaffhausen, Switzerland) was used for additive buildup of specimens. Additive manufacturing of the DMD structures was performed with the integrated laser processing system Ambit S5 from HMT (Hybrid Manufacturing Technologies, Midlands, UK) that includes an IPG YLR-1000-MM-WC (IPG Photonics Corporation, Oxford, MA, USA) fibre laser. The working distance of the processing head was set at 8.0 mm, which resulted in an approximate melt pool width of 2.2 mm. Argon is used as shielding gas, nozzle protection, and powder carrier gas at flow rates of 8 L/min, 6 L/min, and 4 L/min, respectively. The material composition of the used powder is detailed in Table 1.

Table 1. Inconel 718 composition.

Element	Fe	Ni	Cr	Nb + Ta	Mo	Ti	Al	C	N	O
wt.%	18.1	53.4	18.9	5.1	3.05	0.88	0.44	0.05	0.02	0.01

The milling step was carried out with a metalworking fluid (MWF) semi-synthetic water-miscible cooling lubricant Viscotex Zubora 67H Ultra (VISCOTEX Schmiertechnik AG, Altstätten, Switzerland). The selected MWF was recommended for the milling of IN718 by the machine manufacturer. The density of the MWF is 0.97 g/cm^3 at 15 °C. The tests were conducted with 100% lubricant and 5% concentration emulsion.

For the milling tests, a Fraisa W00410404 end mill with W50410013 cutting inserts (Fraisa SA, Bellach, Switzerland) was used. Cutting parameters were chosen in accordance with recommendations from the manufacturer: cutting speed—40 m/min, feed per tooth—0.04 mm, axial infeed depth—1 mm, radial infeed depth—15 mm (30 mm is maximal for this tool), spindle speed—320 revolutions per minute, feed rate—51 mm/min.

2.2. Fabrication of the Specimens

The geometry of the specimens is shown in Figure 1. The scanning raster orientation was changed by 45° for each layer (Figure 1b). The scanning speed was 200 mm/min for the contour and 333 mm/min for the raster [15]. The specimens were produced segment by segment, with intermediate machining between the segments. The segment height is 4 mm, and the whole specimen is 8 mm. These dimensions were chosen to be representative of process conditions in a massive segment. Five sets of three specimens were fabricated for each lubricant concentration with different cleaning techniques applied to the interface between segments (Table 2). Method P1 was chosen as the reference, demonstrating the cladding process on top of the uncleaned and unprocessed DMD part. Method P2 includes manual cleaning with ethanol—a proven, but manual technique. Method P3 includes cleaning with air blasting—a method which can be relatively easily implemented and automatised but might require machine modification. Method P4 includes the laser treatment of the surface with a defocused laser beam with a power of 1000W to evaporate liquid before cladding. Defocusing was carried out by moving the nozzle 10 mm in the +Z direction. This method can be easily implemented by slight modification of G-code and requires no modification of the current setup. Method P5 includes no specific cleaning besides tilting the machine table by 90° for 5 s to let the liquid drop off from the milled surface. This method is also easily implementable and the fastest, but the cleaning quality is reduced to a very low level.

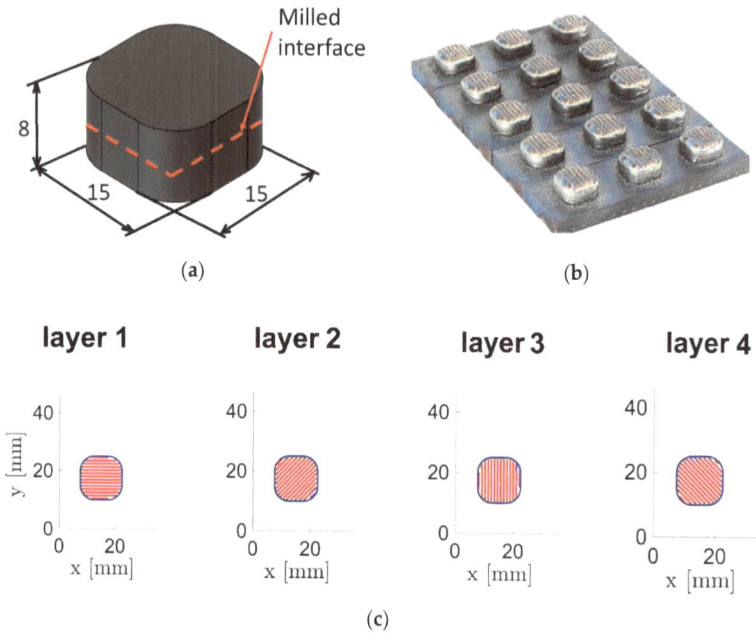

Figure 1. (**a**) Specimen geometry, (**b**) photo of the specimens, and (**c**) raster orientation.

Table 2. Description of the process chain and intermediate cleaning.

Interface Treatment	Fabrication Details			
	1st Segment	Milling	Cleaning	2nd Segment
P1	Cladding	-	-	Cladding
P2		Milling	Ethanol cleaning	
P3			Air Blasting	
P4			Laser Cleaning	
P5			Table Tilting	

This lubricant was applied to the corresponding specimens according to Table 2. Two sets of tests were carried out—with a concentration of 5% of the lubricant in the cutting liquid for the first set and 100% for the second one. These concentrations were chosen as the most commonly used for the machining of IN718. The specimens are named according to the following rule—each sample is named PS_C, where S (from 1 to 5) indicates the process chain and cleaning method according to Table 2, and C is the concentration of the lubricant in the cutting liquid (5 or 100). For example, P5_100 means that 100% concentration of the lubricant in the cutting liquid and the cleaning method P5 were used.

2.3. Experimental Characterisation

The cross sections were ground flat with SiC papers from 240 to 4000 and polished with 1 μm and 0.1 μm diamond suspension. The cross sections were analysed on a Keyence

VHX-1000 digital microscope (Keyence, Osaka, Japan). Kalling II was used as an etchant to reveal the microstructure. In the polished condition, the porosity was determined from ten cross-sections by image analysis with the software ImageJ (version 1.51p, ImageJ, Bethesda, MD, USA).

A Zeiss EVO 10 scanning electron microscope (SEM, Carl Zeiss AG, Jena, Germany) equipped with an Oxford Instruments X-MaxN detector for energy-dispersive X-ray spectroscopy (EDS, Oxford Instruments, Abingdon, UK) was used for SEM and EDS analysis of the specimens. The EDS method can be used for qualitative analyses of light elements (B, C, etc.), although even quantitative analyses can be performed with a proper selection of parameters [20].

2.4. Hydrogen Measurement

Hydrogen content determination was conducted using the hot gas extraction (HGE) method on the LECO ONH 836 (LECO, Ann Arbor, MI, USA) machine. At least three specimens (~0.5 g mass of each specimen) were extracted and measured for each DMD cube for hydrogen determination. The specimens were extracted from the middle of the specimen from the interface zone.

To calculate the hydrogen concentration change in the specimen over time, Fick's diffusion law calculations were performed. Fick's Laws of Diffusion provide the foundational principles for describing the transport of atoms within a medium. Fick's First Law states that the flux of atoms moves from regions of high concentration to low concentration, proportional to the concentration gradient. Mathematically, it is expressed as:

$$J = -D\frac{\partial D}{\partial x}, \tag{1}$$

where J is the diffusion flux, D is the diffusion coefficient, C is the concentration, and x is the spatial coordinate.

Fick's Second Law, which is derived from the first law under the assumption of a constant diffusion coefficient, addresses the change in concentration over time:

$$\frac{\partial C}{\partial t} = D\frac{\partial^2 C}{\partial x^2} \tag{2}$$

This equation describes how diffusion causes the hydrogen concentration to evolve within the Inconel 718 specimen.

The diffusion process was modelled using a Gaussian distribution, centred at the middle of the specimen with an offset of 0.4 mm, to represent the hydrogen concentration profile. This assumption represents the increased H concentration in the layer below the interface. The Gaussian model was chosen for its physical relevance in many diffusive processes.

The concentration profile at any position x and time t was estimated using the equation:

$$C(x,t) = C_{m,init} exp\left(-\frac{(x)^2}{2\sigma_0^2}\right) \tag{3}$$

where $C_{m,init}$ is the peak concentration at the initial time.

The peak concentration was adjusted based on the spread of the distribution to ensure mass conservation:

$$C_{m,init} = C_{m,current} \times \frac{\sigma_{current}}{\sigma_0} \tag{4}$$

where $\sigma_{current}$ is the standard deviation after the selected time, calculated as $\sigma(t) = \sqrt{4Dt}$.

3. Results

An overview of the specimen P2_5 cross-sections is shown in Figure 2. In the figure, the dashed line indicates the interface between the lower and upper sections of the specimen. It can be seen that no additional porosity is observed at the milled interface, except for a minimal number of gas-entrapped pores that occur throughout the specimen due to the non-ideality of the powder material.

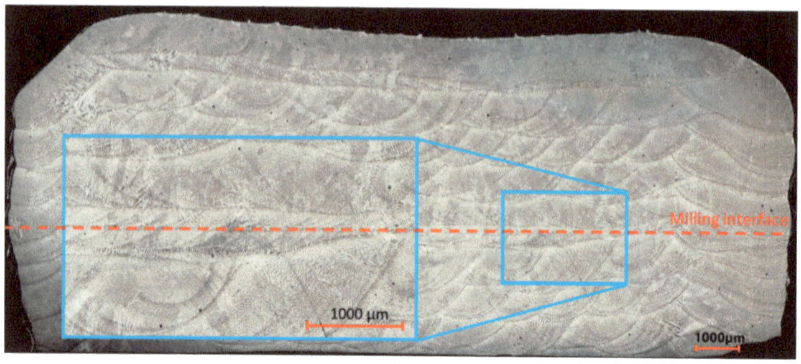

Figure 2. Etched cross-section of the specimen P2_5.

The results of porosity measurement by the optical method are presented in Figure 3. The porosity was measured on the polished specimens in three areas of the interface for each specimen. The measurements took place no further than one layer away from the interface. The porosity result for a DMD specimen, fabricated without interim milling, is 0.1%. The results of specimen P1_5 show porosity values comparable to P5_5. The porosity level of the specimen P2_5 was chosen as a base level due to the most careful cleaning being applied. The specimen P1 shows higher porosity values in comparison with P2, P3, and P4 due to the non-smooth and oxidized surface generated as the last layer by DMD in P1 compared to the milled surfaces in the other process chains. The results for the three applied cleaning techniques (P2, P3, P4) show very close results at approximately 0.1%, not depending on the cutting liquid concentration. The specimen P5_100 shows the highest porosity for these tests, approximately 0.3%.

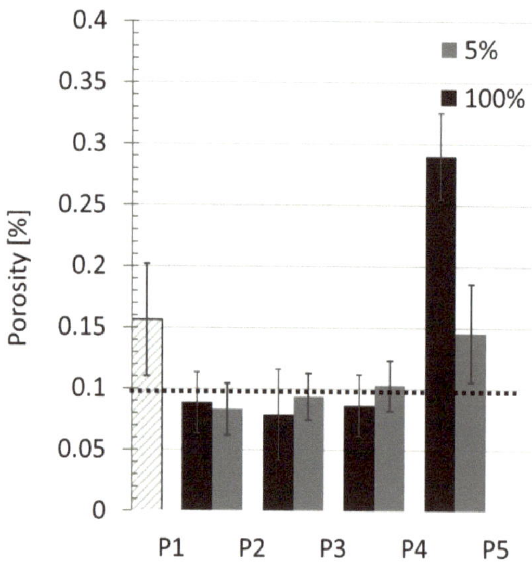

Figure 3. Porosity levels for the obtained specimens from the five different process chains according to Table 2 and two different concentrations of the lubricant. Dotted line shows the target value—DMD-fabricated specimen with no interim process interruption. P1 is marked with a different pattern since it does not refer to any cleaning technique and liquid concentration.

The hot gas extraction method was used to measure the hydrogen content, and the most representative specimens P1, P2, and P5 were selected for analysis. The measurement results are shown in Figure 4. The graph shows that the hydrogen content in the interface of samples P1, P2, and P5_100 is below 2 ppm. Sample P5_5, on the other hand, shows an average value slightly higher than 2 ppm.

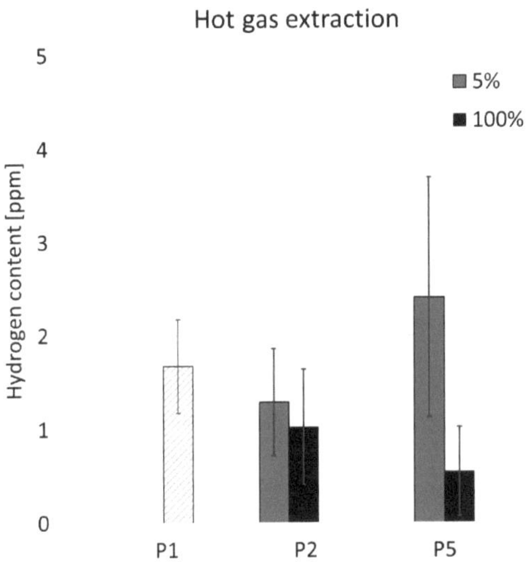

Figure 4. Hydrogen content of the investigated specimens, measured by the hot gas extraction method. P1 is marked with a different pattern since it does not refer to any cleaning technique and liquid concentration.

The hydrogen concentration in the specimens was measured 60 days after they were fabricated. They were stored at 20 °C and cut right before the HGE. Therefore, it was assumed that the concentration in the specimen interface decreased due to diffusion. To calculate the change in hydrogen concentration in the specimen with time, Fick's diffusion law calculations were carried out. The diffusion coefficient D was assumed to be 5×10^{-15} m^2/s, according to Aiello et al. [19]. The average value for the P5_5 sample, which is 2 ppm, was used in the calculations. The equations used were described in Section 2, and the results are shown in Figure 5.

It can be seen that the width of the distribution increased over time, and the peak concentration decreased. The calculation showed that immediately after fabrication the concentration was 3.2 ppm. This change is relatively small for such a time interval due to the low Inconel diffusion coefficient and the absence of elevated temperature.

Also, by extracting the HGE sample from the middle of the DMD sample, the capture of a small part of the material not exposed to hydrogen can be assumed, which might have affected the measurement results to a small extent, reducing them. Therefore, the authors estimate that the real concentrations in the interface zone are slightly higher.

The results of carbon content measurement by EDS are shown in Figure 6. Four lines were measured in the interface area for each specimen. A secondary electron (SE) image with the marked milling interface and measured carbon content lines is shown in Figure 6a. Examples of the obtained linear element scans are shown in Figure 6b. None of the measured specimens show any detectable increase in carbon content in the observed interface.

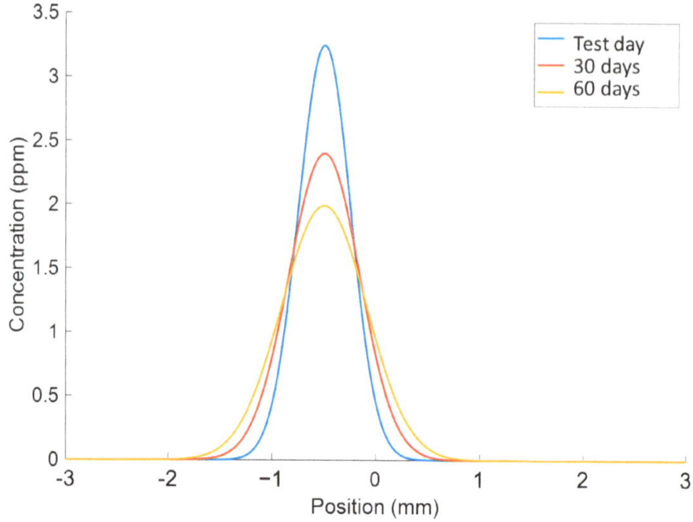

Figure 5. Evolution of hydrogen concentration in IN718 (P5_5).

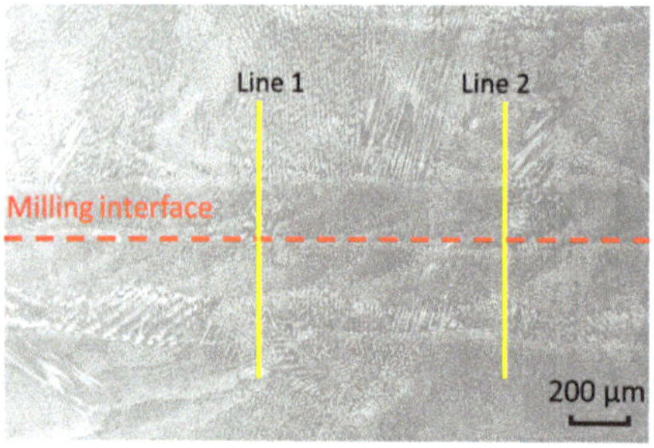

(**a**)

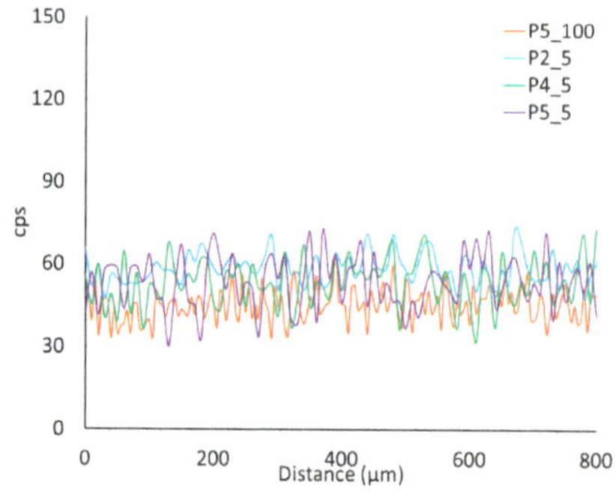

(**b**)

Figure 6. Carbon content determination. (**a**) SE image of the interface P2_5. (**b**) Comparison of representative linear measurements for the listed specimens.

4. Discussion

An important parameter in this process is the temperature after cladding, before milling, and after milling, which can affect the porosity and microstructure of the parts. The specimen surface temperature after cladding was similar for all specimens due to the same process parameters and, therefore, was not considered. After cladding, there was a dwell time applied for about 15 min for technological reasons (tool change, part height measurement, etc.), when the specimens cooled down to temperatures below 50 °C. Also, a coolant was used during the milling process which ensured that the temperature was reduced without further affecting the material structure. This also ensured that the temperature of the specimen surface after milling and before the start of the cladding of the next segment was in the ambient temperature range.

Regarding the laser cleaning method, it may be noted that laser surface treatment involves high-temperature thermal processing, which can critically impact the final product quality. High temperatures can lead to thermal erosion and surface imperfections, such as micro-cracks and cavities, limiting the practical applications of laser processing [21]. However, in this work, the laser parameters were chosen to minimize the thermal effects but, at the same time, effectively remove the liquid from the investigated area. No remelted layer or defects due to laser influence are observed on cross-sections of P4 specimens.

It is important to control the porosity level since it has a negative effect on the mechanical properties [22,23]. According to the porosity analysis, it can be concluded that the cutting liquid has a negative influence on the quality of the interface in the segmental cladding. The reference specimen P1 shows higher porosity values, probably due to the fact that the surface is rough (typical for the DMD process). The air or laser-assisted cleaning methods showed almost the same effectiveness in reducing porosity. The specimen with ethanol cleaning showed the lowest porosity value. The specimen P5 shows the highest porosity values. Porosity is probably formed due to the vaporising liquid bubbles trapped in the molten material. Therefore, the different cleaning methods showed comparable effectiveness as they removed cutting liquid from the surface.

Analysis of carbon content did not detect increased values at the interface in any of the samples studied. The assumption of carbon saturation of the alloy is related to the composition of the lubricant—consisting of more than 80% carbon. These results may be due to the fact that the lubricant present in the cutting fluid evaporates in the process faster than the diffusion of carbon into the alloy.

For high-performance materials like Inconel 718, manufacturers strive to reduce hydrogen levels to below 1 ppm if possible, especially for critical applications [19]. Control of the hydrogen level is important due to the risk of hydrogen embrittlement [24,25]. The hydrogen content analysis showed that the hydrogen concentration in the specimens is slightly elevated for this material. Ideally, keeping the hydrogen content as low as possible minimises the risk of hydrogen embrittlement. Sample P5_5 showed an increased hydrogen content of up to 3 ppm, which, in accordance with the data from the research [10], can lead to a decrease in fracture strain from 40% to 30% being attributed to hydrogen embrittlement. The elevated hydrogen content in specimens produced with 5% lubricant at the interface could be due to the vaporization and dissociation of water and oil in the plasma created by the laser beam. Hydrogen is then resolved in the metal melt pool.

These defects can seriously affect the strength and durability of the material, especially in critical applications where high reliability is required. The importance of controlling the hydrogen content cannot be overestimated, as even a slight increase in the content of this element leads to a significant reduction in the ductility of the alloy through the mechanism of hydrogen embrittlement. The results of the study emphasise the need for the implementation of cleaning methods prior to cladding in the studied process to reduce porosity and hydrogen uptake.

5. Conclusions

The results obtained allow a comparison of segment-by-segment cladding without and with the use of milling between segments, as well as different methods of cleaning the obtained interface. The following conclusions can be drawn:

- It is important to perform the top milling of the "base" segment in a segment-by-segment DMD process as it flattens the surface, resulting in a higher interface quality;
- In terms of porosity, tilting of the machine table has been shown to be the least effective compared to all the other applied methods. Air blasting and laser-assisted cleaning methods have proven effective in reducing porosity and can be relatively easily implemented in the existing process chain as it can be automated. Ethanol cleaning has shown the best results in this respect but is the most economically unfavourable solution due to the difficulties in automation and implementation in a continuous process;
- Cleaning the part surface is necessary not only to reduce porosity but also to reduce hydrogen saturation on the surface, which can lead to hydrogen embrittlement. This effect is more pronounced with an emulsion than with a pure lubricant because of the significant influence of water in this liquid.

Thus, automated cleaning methods, such as laser or air blasting, are generally sufficient and help avoid the constraints of dry cutting, which is particularly disadvantageous for hard-to-machine materials like nickel-based superalloys. However, for critical parts, manual cleaning with ethanol is advised to enhance interface quality, despite a reduction in productivity. This approach ensures the reliability needed for such components, while less critical applications may benefit from the efficiency of automated techniques.

Author Contributions: Conceptualization, S.E. and K.W.; methodology, S.E.; software, S.E.; validation, S.E. and K.W.; formal analysis, S.E.; investigation, S.E.; resources, T.S.; data curation, S.E.; writing—original draft preparation, S.E.; writing—review and editing, K.W.; visualization, S.E.; supervision, K.W. and T.S.; project administration, S.E. and T.S.; funding acquisition, K.W. and T.S. All authors have read and agreed to the published version of the manuscript.

Funding: The authors would like to acknowledge the support of the funding agency Innosuisse (grant number 25498) and the companies GF Machining Solutions, Fraisa SA, EMPA and Accelleron Turbo Systems Switzerland Ltd.

Institutional Review Board Statement: Not applicable.

Informed Consent Statement: Not applicable.

Data Availability Statement: The datasets generated during and/or analysed during the current study are available from the corresponding author on reasonable request.

Acknowledgments: The authors would like to thank Knut Krieger for his assistance with this study.

Conflicts of Interest: All authors were employed by the company Inspire AG. The authors declare that the research was conducted in the absence of any commercial or financial relationships that could be construed as a potential conflict of interest.

References

1. *ASTM F3055-14a*; Standard Specification for Additive Manufacturing Nickel Alloy (UNS N07718) with Powder Bed Fusion. ASTM International: West Conshohocken, PA, USA, 2021.
2. Tucho, W.M.; Cuvillier, P.; Sjolyst-Kverneland, A.; Hansen, V. Microstructure and hardness studies of Inconel 718 manufactured by selective laser melting before and after solution heat treatment. *Mater. Sci. Eng. A* **2017**, *689*, 220–232. [CrossRef]
3. Zhang, D.; Feng, Z.; Wang, C.; Wang, W.; Liu, Z.; Niu, W. Comparison of microstructures and mechanical properties of Inconel 718 alloy processed by selective laser melting and casting. *Mater. Sci. Eng. A* **2018**, *724*, 357–367. [CrossRef]
4. Bhatt, A.; Attia, H.; Vargas, R.; Thomson, V. Wear mechanisms of WC coated and uncoated tools in finish turning of Inconel 718. *Tribol. Int.* **2010**, *43*, 1113–1121. [CrossRef]
5. Obikawa, T.; Yamaguchi, M.; Funai, K.; Kamata, Y.; Yamada, S. Air jet assisted machining of nickel-base superalloy. *Int. J. Mach. Tools Manuf.* **2012**, *61*, 20–26. [CrossRef]
6. Devillez, A.; Schneider, F.; Dominiak, S.; Dudzinski, D.; Larrouquere, D. Cutting forces and wear in dry machining of Inconel 718 with coated carbide tools. *Wear* **2007**, *262*, 931–942. [CrossRef]
7. Lei, X.; Wei, X.; He, Y. Effect of cooling and lubrication conditions on cutting performance and surface integrity of Inconel 718 superalloy in end face milling. *Adv. Mech. Eng.* **2023**, *15*, 16878132231158217. [CrossRef]
8. Ezugwu, E.O.; Bonney, J.; Olajire, K.A. The effect of coolant concentration on the machinability of nickel-base, nimonic C-263, alloy. *Tribol. Lett.* **2004**, *16*, 311–316. [CrossRef]
9. Liang, S.Y.; Feng, Y.; Ning, J. Predictive Manufacturing: Subtractive and Additive. In *IOP Conference Series: Materials Science and Engineering*; IOP Publishing: Bristol, UK, 2020; Volume 842, No. 1; p. 012024.
10. Careri, F.; Imbrogno, S.; Umbrello, D.; Outeiro, J.C.; Batista, A.C. A residual stress prediction of machining IN718 produced by direct energy deposition. *Procedia CIRP* **2021**, *102*, 13–18. [CrossRef]
11. Quadra Vieira dos Santos, G.; Kaneko, J.I.; Abe, T. Analysis of Machinability on Properties of Inconel 718 Wire and Arc Additive Manufacturing Products. *J. Manuf. Mater. Process.* **2023**, *8*, 4. [CrossRef]
12. Holmberg, J.; Berglund, J.; Brohede, U.; Åkerfeldt, P.; Sandell, V.; Rashid, A.; Hosseini, S. Machining of additively manufactured alloy 718 in as-built and heat-treated condition: Surface integrity and cutting tool wear. *Int. J. Adv. Manuf. Technol.* **2024**, *130*, 1823–1842. [CrossRef]
13. Schneider, S.A.; Hermsen, S.; Kirchmann, S.; Ganser, P.; Bergs, T.; Schleifenbaum, J.H. Machinability analysis for milling of additively manufactured Inconel 718 with specifically induced porosity. In Proceedings of the Machining Innovations Conference for Aerospace Industry (MIC), Semarang, Indonesia, 12 November 2022.
14. Song, Y.; Park, S.; Choi, D.; Jee, H. 3D welding and milling: Part I-a direct approach for freeform fabrication of metallic prototypes. *Int. J. Mach. Tools Manuf.* **2005**, *45*, 1057–1062. [CrossRef]

15. Soffel, F.; Eisenbarth, D.; Hosseini, E.; Wegener, K. Interface strength and mechanical properties of Inconel 718 processed sequentially by casting, milling, and direct metal deposition. *J. Mater. Process. Technol.* **2021**, *291*, 117021. [CrossRef]
16. Barckhoff, J.R. *Total Welding Management*; American Welding Society: Doral, FL, USA, 2005.
17. AlShaer, A.W.; Li, L.; Mistry, A. The effects of short pulse laser surface cleaning on porosity formation and reduction in laser welding of aluminium alloy for automotive component manufacture. *Opt. Laser Technol.* **2014**, *64*, 162–171. [CrossRef]
18. Cortina, M.; Arrizubieta, J.I.; Ukar, E.; Lamikiz, A. Analysis of the influence of the use of cutting fluid in hybrid processes of machining and laser metal deposition (LMD). *Coatings* **2018**, *8*, 61. [CrossRef]
19. Aiello, F.; Beghini, M.; Bertini, L.; Macoretta, G.; Monelli, B.D.; Valentini, R. Hydrogen diffusivity and tensile properties degradation in SLMed Inconel 718. In *IOP Conference Series: Materials Science and Engineering*; IOP Publishing: Bristol, UK, 2022; Volume 1214, No. 1; p. 012002.
20. Wu, W.; Liu, Z.W.; Lin, C.C.; Hua, J.J.; Zeng, Y. Application of Low Voltage in Quantitative Analysis by Energy Dispersive Spectrum (EDS). In *Materials Science Forum*; Trans Tech Publications, Ltd.: Bäch, Switzerland, 2014; Volume 804, pp. 165–168. [CrossRef]
21. Yilbas, B.S.; Ali, H.; Karatas, C.; Al-Sharafi, A. Laser texturing of Inconel 718 alloy surface: Influence of environmental dust in humid air ambient. *Opt. Laser Technol.* **2018**, *108*, 346–354. [CrossRef]
22. Kan, W.H.; Chiu LN, S.; Lim CV, S.; Zhu, Y.; Tian, Y.; Jiang, D.; Huang, A. A critical review on the effects of process-induced porosity on the mechanical properties of alloys fabricated by laser powder bed fusion. *J. Mater. Sci.* **2022**, *57*, 9818–9865. [CrossRef]
23. Al-Maharma, A.Y.; Patil, S.P.; Markert, B. Effects of porosity on the mechanical properties of additively manufactured components: A critical review. *Mater. Res. Express* **2020**, *7*, 122001. [CrossRef]
24. Li, X.; Zhang, J.; Akiyama, E.; Fu, Q.; Li, Q. Hydrogen embrittlement behavior of Inconel 718 alloy at room temperature. *J. Mater. Sci. Technol.* **2019**, *35*, 499–502. [CrossRef]
25. Xu, J.; Hao, Z.; Fu, Z.; He, X.; Wang, H.; Xu, G. Hydrogen embrittlement behavior of selective laser-melted Inconel 718 alloy. *J. Mater. Res. Technol.* **2023**, *23*, 359–369. [CrossRef]

materials

MDPI

Article

Accuracy of Mandibular Removable Partial Denture Frameworks Fabricated by 3D Printing and Conventional Techniques

Soonam Kim [1,†], Kyung Chul Oh [2,†] and Jee-Hwan Kim [2,*]

1 Department of Dentistry, Graduate School, Yonsei University, Seoul 03722, Republic of Korea; snkim2016@gmail.com
2 Department of Prosthodontics, College of Dentistry, Yonsei University, Seoul 03722, Republic of Korea; kyungabc@yuhs.ac
* Correspondence: jee917@yuhs.ac; Tel.: +82-2-22283161; Fax: +82-2-3123598
† These authors contributed equally to this work.

Abstract: Herein, we used digital superimposition to evaluate the accuracy of metal frameworks for mandibular removable partial dentures fabricated using three techniques. Thirty master casts of a mandibular dentiform were categorized into three groups ($n = 10$) based on the framework manufacturing method: selective laser melting-based metal three-dimensional (3D) printing (SLM), digital light projection-based resin 3D printing and subsequent casting (RPC), and conventional casting (CON). The master casts were scanned twice, initially after preparation and subsequently after attaching silicone using the frameworks. These scan files were digitally superimposed to measure the silicone thickness. Statistical analysis was conducted using SPSS Statistics (Version 23.0, IBM Corp., Somers, NY, USA). One-way ANOVA and a post hoc Tukey's multiple comparison tests were performed to determine differences among the three groups ($\alpha = 0.05$). The RPC group exhibited significantly higher overall and mean internal discrepancies at rest and tissue stops than the SLM and CON groups, which exhibited statistically insignificant differences. Thus, SLM fabrication resulted in comparable accuracy to that achieved by CON, whereas sequentially performing resin 3D printing and casting induced inferior accuracy. However, all frameworks across the three groups were clinically acceptable.

Keywords: accuracy; removable partial denture framework; selective laser melting; superimposition; 3D printing

Citation: Kim, S.; Oh, K.C.; Kim, J.-H. Accuracy of Mandibular Removable Partial Denture Frameworks Fabricated by 3D Printing and Conventional Techniques. *Materials* **2024**, *17*, 3148. https://doi.org/10.3390/ma17133148

Academic Editors: Nikolaos Kladovasilakis, Konstantinos Tsongas and Dimitrios Tzetzis

Received: 27 May 2024
Revised: 21 June 2024
Accepted: 25 June 2024
Published: 27 June 2024

1. Introduction

Manufacturing methods for the metal frameworks of removable partial dentures (RPDs) have recently transitioned from conventional lost-wax techniques to computer-aided design (CAD) and computer-aided manufacturing (CAM) techniques, specifically metal three-dimensional (3D) printing. Considering the history of the manufacturing methods of RPD metal frameworks, the transition to the current direct 3D-printed RPD frameworks provides an innovative path. Fauchard first reported the use of metal structures in an RPD in 1728, whereby he used metal labial and lingual bars to connect two carved ivory blocks of the designed RPD [1,2]. In the late 1890s, over 160 years later, the lost-wax casting technique was introduced in dentistry to fabricate metal prostheses, including RPD metal frameworks. However, it comprised complex procedures, the making of a refractory cast, waxing, investing, and casting, thereby requiring a large amount of labor and time [3–5]. Moreover, the distortion of the wax pattern on a refractory cast caused inaccuracies in the RPD frameworks, which can hinder the success of the RPD treatment [6]. Despite these drawbacks, conventional casting techniques were used for over a century until CAD/CAM technology was introduced in dentistry in the 1970s, bringing revolutionary changes in the manufacturing methods for all components from inlays to RPD metal frameworks [7–11].

Early CAD/CAM methods involved milling, which is a subtractive technique. However, this technique had several disadvantages, such as the wear of cutting tools, complicated shapes or undercut areas, wastage of cutting chips, long processing time, and shrinkage [12,13]. As such, 3D printing technology has replaced the milling method as it bypasses these limitations. It is an additive manufacturing (AM) technique that produces 3D objects by adding materials in layers, rendering it highly efficient in manufacturing complex-shaped objects [4]. Currently, several 3D printing technologies, such as stereolithography (SLA), digital light projection (DLP), selective laser sintering (SLS), and selective laser melting (SLM), are used depending on the material and energy sources. In this study, SLM and DLP were used [9–11,14].

SLA was the first 3D printing technique patented in the 1980s and has been used to manufacture various objects. It involves using ultraviolet lasers to polymerize resin in layer thicknesses ranging from 10 to 100 μm [15,16]. DLP, which was invented by Larry Hornbeck of Texas Instruments in 1987 [17], has a resolution and use range similar to that of SLA. In fact, DLP and SLA are considered to be in the same AM category by the American Section of the International Association for Testing Materials [17]. However, the primary difference between these methods is the light source, that is, the image is formed by an arc lamp or micromirrors, respectively. DLP has a shorter processing time than SLA as the projecting light can cure an entire layer at once [16,18]. The number of micromirrors determines the resolution of the projected image in DLP [17,19]. SLA was first used to print resin sacrificial patterns for RPD frameworks from 3D CAD models in the early 2000s; however, the method required casting [20]. Later, the direct printing of metal RPD frameworks using an SLM 3D printer was achieved, thereby eliminating the complex conventional casting processes, except for finishing and polishing [21,22].

SLM involves melting metal powder using a high-energy laser beam, fusing, and solidifying it in layers according to the CAD information [14,16]. SLM and SLS principles are similar; however, the primary difference between the two methods is the material used [14], that is, SLS and SLM are preferred for ceramics/polymers and metals, respectively. Meanwhile, Suzuki et al. [13] stated that various materials, including metals, could be used in SLS, and the difference between SLS and SLM is the method of manipulating the powder. In particular, SLM involves melting the powder, whereas SLS entails sintering it. Alageel et al. [22] stated that SLM involves the full melting of metal powder, whereas SLS involves its partial melting. In this study, the term SLM implies the metal 3D printing technique.

The use of SLM-fabricated RPD frameworks has increased recently; however, conventional casting techniques remain time- and labor-intensive. To generalize the use of convenient metal 3D-printed frameworks, the advantageous properties of 3D-printed chrome–cobalt (Co–Cr) alloy should be proven, and the accuracy of the frameworks should be comparable to that of conventional cast frameworks. Several studies have reported that SLM-fabricated Co–Cr alloys have more homogeneous microstructures than cast Co–Cr alloys, resulting in enhanced mechanical properties [14,23–27]. For the accuracy of metal 3D-printed RPD frameworks, recent review articles concluded that RPD frameworks fabricated with SLM and conventional casting exhibited similar accuracy within the clinically acceptable range [13,27–29]. However, certain studies have noted that although the accuracy values are clinically acceptable, the internal discrepancies of SLM-fabricated RPD frameworks are larger than those of conventionally cast frameworks [30–32]. Moreover, the studies on mandibular 3D-printed RPD frameworks are limited. Mandibular RPD frameworks are u-shaped and have a significantly smaller contact area with tissue than maxillary RPD frameworks owing to the absence of a palatal area, where most of the contact with maxillary RPD frameworks is concentrated, which might affect the internal adaptation of the components of RPD frameworks. In addition, despite the various methods employed to investigate the accuracy of RPD frameworks manufactured through CAD/CAM techniques [30–37], comparative studies on the three types of fabrication methods using reliable digital measurements are rare. Previous research predominantly compared metal

measurement step (Section 2.4). Each RPD framework was fitted onto the corresponding master cast using silicone. Before fitting, the intaglio surface of the framework was coated with a thin layer of petroleum jelly (Vaseline, Unilever, Greenwich, CT, USA) to prevent the silicone from attaching to the framework. Additionally, a thin adhesive liquid (3M ESPE Polyether Adhesive, 3M ESPE AG, Bayern, Germany) was applied to the measuring areas of the cast to prevent detachment of the silicone during framework removal. Mixed vinyl polyether silicone (Fit Checker Advanced; GC Corp., Tokyo, Japan) was applied to the intaglio surface of the rests, tissue stops, and lingual bars. The framework was immediately fitted onto the corresponding cast and carefully removed after the silicone hardened, leaving the silicone intact on the casts. If any gap was detected between the silicone and the measuring area of the cast, the process was repeated after cleaning the cast. Each master cast with silicone was rescanned using the tabletop scanner. All scanning procedures were performed without a powder coating.

2.4. Measurement of the Internal Discrepancies of the RPD Frameworks

The measurement steps using the 3D metrology software are presented in Figure 3. The two types of STL files, namely that of the master cast only and the silicone-attached master cast, were superimposed using a local best-fit alignment function of the metrology software (GOM Inspect 2018, Carl Zeiss GOM Metrology GmbH, Braunschweig, Germany) to measure the thickness of the silicone, representing the internal discrepancy of the framework (Figure 3A). The mean deviation values in the eight areas of the RPD framework were measured: three rests (35R, 44R, and 47R), four tissue stops (36T, 37T, 45T, and 46T), and a lingual bar. The borderlines of the measuring areas were manually created in each master cast following a certain standard to ensure designation consistency. Each borderline was chosen to be slightly away from the edge, and the perpendicular borders of lingual bars were selected at the distal sides of the canines (Figure 3B). Each selected patch was inspected using the "surface comparison on actual" function of the metrology software and visualized using color mapping and tabulating the mean and maximum deviation values (Figure 3C).

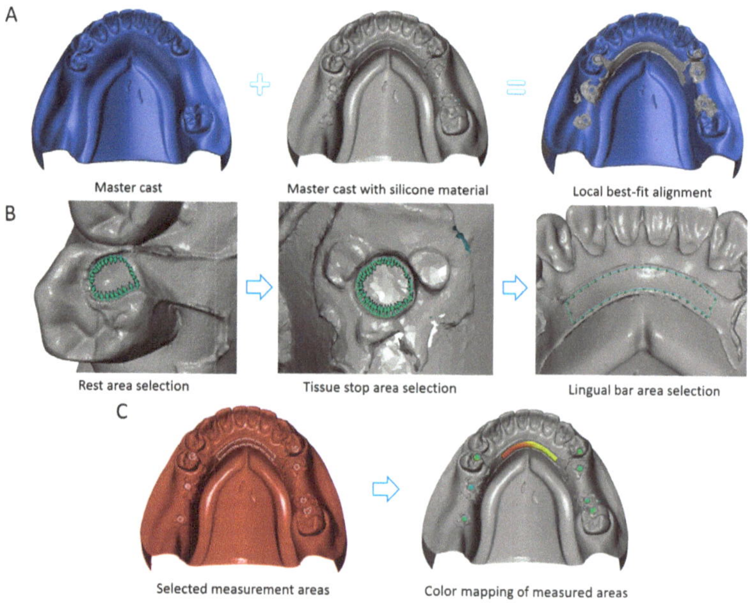

Figure 3. Screenshots of the measurement steps with the 3D metrology software. The thickness of the imprinted silicone material representing the internal discrepancies was measured by superimposing

two STL files: the cast only and the silicone-attached cast. (**A**) Superimposition of the two STL files by local best-fit alignment function of the metrology software, (**B**) manual selection of each measurement area, and (**C**) selected border lines at eight measurement areas (three rests, four tissue stops, and one lingual bar area) and color mapping of the areas (green represents good fit, yellow to red represents positive error, blue represents negative error).

The internal discrepancies at the rests, tissue stops, and lingual bars (IDR, IDT, and IDL, respectively) were measured. In addition, the overall internal discrepancy (IDO), comprising the average values of the IDR and IDT, was calculated.

2.5. Statistical Analysis

The sample size of 10 per group was determined based on the statistical significance level set at $\alpha = 0.05$ and power $(1 - \beta)$ of 0.8 with an effect size of 0.6. The Shapiro–Wilk test was performed to examine the normality and all data followed normal distribution ($p > 0.05$). One-way analysis of variance and a post hoc Tukey's multiple comparison test were performed to determine differences between the three groups ($\alpha = 0.05$). The data were analyzed using statistical software (SPSS Statistics version 23.0, IBM Corp., Somers, NY, USA), and the analysis graphs were created using the graphing software (GraphPad Prism 10, Boston, MA, USA).

3. Results

The overall internal discrepancies (IDOs) and internal discrepancies at rests, tissue stops, and lingual bars (IDR, IDT, and IDL, respectively) of mandibular RPD metal frameworks fabricated using three methods are presented in Figure 4. The RPC group shows significantly higher IDOs than the SLM and CON groups ($p = 0.001$ and 0.019, respectively), whereas the IDOs of the SLM and CON groups do not differ significantly ($p = 0.633$). The CON group has the lowest mean IDR ($p = 0.010$, $p < 0.001$), whereas the SLM group has the lowest IDT ($p = 0.001$, $p = 0.025$). The RPC group, combined with resin 3D printing and casting, shows the highest IDRs and IDTs with significant differences. For the IDLs, no significant differences are noted among the three groups. The mean IDRs, IDTs, and IDLs of the mandibular RPD frameworks in the three groups are presented in Table 1.

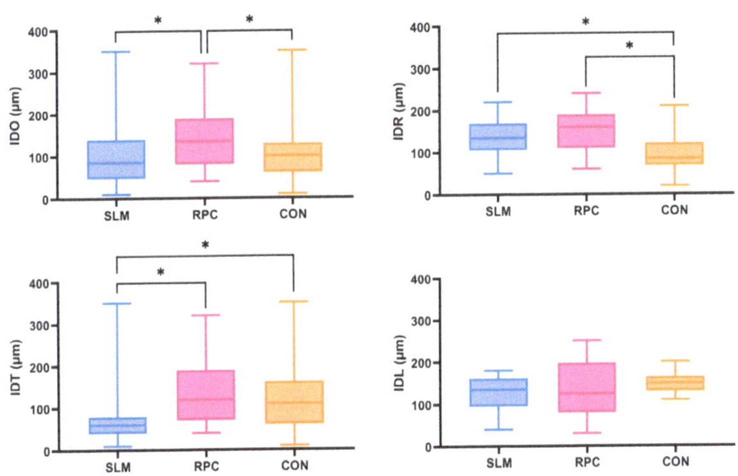

Figure 4. Comparisons of the overall internal discrepancies (IDOs) and internal discrepancies at rests, tissue stops, and lingual bars (IDR, IDT, and IDL, respectively) of mandibular RPD metal frameworks fabricated using three methods (SLM: selective laser melting-based metal 3D printing, RPC: DLP-based resin 3D printing and subsequent casting, CON: conventional lost-wax casting). The asterisks indicate statistically significant differences among the three groups ($p < 0.05$).

Table 1. Internal discrepancies (μm) of the mandibular RPD metal frameworks fabricated using three methods.

Internal Gap/Group	SLM	RPC	CON
IDO	101.7 ± 68.41 [a] (81.48–121.95)	143.7 ± 67.29 [b] (123.43–163.99)	112.3 ± 70.09 [a] (88.83–135.74)
IDR	133.0 ± 43.87 [a] (120.92–145.08)	149.0 ± 50.33 [a] (127.09–170.91)	96.3 ± 48.74 [b] (78.02–114.65)
IDT	78.2 ± 74.38 [a] (49.23–107.27)	139.8 ± 78.03 [b] (114.02–165.48)	124.3 ± 81.14 [b] (88.66–159.84)
IDL	127.0 ± 44.48 [a] (95.18–158.82)	135.0 ± 72.91 [a] (82.84–187.16)	150.0 ± 25.81 [a] (131.53–168.47)

The data are expressed as mean ± standard deviation and confidence intervals. IDO, overall internal discrepancy; IDR, internal discrepancy of rests; IDT, internal discrepancy of tissue stops; IDL, internal discrepancy of the lingual bar; SLM group, metal 3D-printed frameworks; RPC group, 3D-printed resin-cast frameworks; CON group, conventional cast frameworks. The default value of 200 μm was subtracted from IDL in SLM and RPC groups. The different letters indicate statistically significant differences among the three groups ($p < 0.05$).

4. Discussion

The null hypothesis was rejected owing to significant differences in the accuracy of the mandibular RPD metal frameworks fabricated using the three manufacturing methods. There were no significant differences between the IDOs of the SLM and CON groups, whereas the RPC group showed a higher IDO than the other two groups, which can be ascribed to the higher error tendency of the RPD frameworks of the RPC group. Although the RPC group eliminated the steps of fabricating a refractory cast and waxing, which could result in inaccuracies owing to the physical properties of wax [6], it was implicated in the errors from the steps of 3D printing of castable resin patterns and in those from the steps of conventional investing and casting. According to Revilla-Leon and Ozcan [17], discrepancies can be incorporated into each step of a digital dental workflow. The 3D printer parameters, the material used (which has its optimal activation range of wavelength), power, and exposition time for AM on the 3D printers can affect the accuracy of the printed objects [17].

The accuracy of the rests and tissue stops, which are structural components of the RPD frameworks that directly contact the tooth or tissue, differed significantly among the three groups. The CON group showed the highest accuracy for the rests, and the SLM group had the highest accuracy for tissue stops. The RPC group had the lowest accuracy for both components. In 2020, Tasaka et al. [36] reported that the accuracy of mandibular RPD metal frameworks differed depending on the structural component comparing SLS technology-based 3D printing and 3D-printed resin pattern casting, which is consistent with the results of this study. In 2019, Bajunaid et al. [37] compared the accuracy of mandibular RPD frameworks fabricated by SLM technology-based 3D printing and conventional casting through measuring four rest zones with a digital microscope. The zones with the highest fit and accuracy among the two groups differed, which is also consistent with this study.

As this study was conducted on the mandibular RPD frameworks, it was also compared with previous studies on the maxillary RPD frameworks. Oh et al. [35] compared the accuracy of maxillary RPD metal frameworks under similar conditions with this study, concluding that there were no significant differences among the three groups of SLM, RPC, and CON; however, the IDOs (226.99–365.30 μm) were all higher than those of this study (101.70–143.70 μm). The mean IDR of the three groups in this study was also lower by approximately 110 μm than that of the study of Oh et al. [35]. This difference could be ascribed to different factors. First, the palatal contact area was not included in the mandibular RPD frameworks, which can reduce the interferences before the contact of rests and rest seats. In contrast, the maxillary RPD frameworks could have early interferences due to the palatal contact of a major connector. Second, the measurement criteria differed. "Point" measurement was employed in the previous study, whereas "area" measurement was employed in this study. The "area" measurement can reduce the contingency of manual designation more than the "point" measurement since it includes countless points within the designated border lines. Lastly, the use of equipment and metrology software from different manufacturing companies could affect the difference. Chen et al. [38] evaluated the adaptation of maxillary RPD metal frameworks fabricated by the SLM technique with four types of partially edentulous resin models. They reported that SLM-fabricated RPD frameworks had acceptable accuracies; however, among the frameworks with a large span and more retainers and clasps, the conventional casting technique exhibited slightly better fit and accuracy. Moreover, Soltanzadeh et al. [30] evaluated the accuracy and fit of maxillary RPD frameworks fabricated by conventional casting and 3D printing techniques with stone and 3D-printed resin models. Both methods revealed clinically acceptable adaptation (50–311 μm), but the conventional casting groups exhibited better overall fit and higher accuracy. The poorest fit was observed at the anterior palatal straps fabricated using the 3D printing technique.

Studies on 3D-printed RPD metal frameworks regardless of maxilla and mandible have previously been conducted. Tregerman et al. [16] compared the clinical fit of RPD metal frameworks fabricated by three workflows—conventional casting pathway, SLM 3D printing with extraoral scanning of the stone cast, and SLM 3D printing with intraoral scanning—and concluded that the completely digital workflow had the lowest misfit. Almufleh et al. [39] compared patient satisfaction with RPDs using frameworks fabricated by conventional casting and SLS 3D printing, revealing that higher satisfaction was achieved with the RPDs obtained using SLS 3D-printed frameworks. Peng et al. [40] compared the trueness of RPD metal frameworks fabricated by SLM and 3D-printed resin-casting. The frameworks fabricated by SLM 3D printing exhibited higher trueness than those by the combined method. Summarizing the results of recent studies, SLM 3D printing and conventional casting techniques demonstrated similar accuracies for fabricating RPD metal frameworks within a clinically acceptable range, as evidenced in this study.

For the internal gap between a rest and a rest seat, the mean distance per rest was reported as 69–387 μm [41] and 193–203 μm [42]. Lee et al. [34] studied the accuracy of RPD frameworks fabricated through a combined method. The mean IDR of 249.27 ± 134.84 μm, which was higher than previously reported values on casted RPD frameworks, was obtained. Oh et al. [35] obtained IDRs in the range of 211.91 ± 16.84 to 259.26 ± 45.41 μm without significant differences among the three manufacturing methods. Souza Curinga et al. [27] achieved the IDR range of 20–279 μm for conventional cast frameworks and 30–272 μm for 3D-printed frameworks, which did not indicate a significant difference. The IDRs in the aforementioned studies are in the range of 20–387 μm, whereas those in this study were lower, indicating clinically acceptable values.

The internal discrepancies of lingual bars, which have not been previously investigated, were also measured in this study, whereby no significant differences were noted among the three groups. The default value of 200 μm was set for lingual relief in CAD subtracted from the IDLs of the SLM and RPC groups because wax relief under a lingual bar was not performed in the CON group owing to the absence of undercuts. We also tried to identify specific tendencies depending on the locational factors among three groups, such as the tooth-borne area versus tooth and mucosa-borne area; however, it was difficult to observe particular trends. The table and figures of the additional regional analysis are included in the Supplementary Materials (Supplemental Table S1, Figures S1 and S2).

Nonetheless, this study had certain limitations. First, this work was an in vitro study with different conditions than that with an actual patient's oral mucosa. The edentulous area of a patient is covered with elastic soft tissue and saliva, whereas the master cast is not. Thus, there might be differences compared to results obtained from in vivo studies. Second, finishing and polishing can have various influences on the results. However, the complete adaptation of the RPD framework on the cast is difficult to achieve without finishing and polishing, and an actual RPD framework is fitted into the patient's oral cavity after complete polishing. Therefore, the measurement of polished frameworks was considered inevitable and appropriate. However, all 30 frameworks were finished and polished by a single experienced board-certified laboratory technician to minimize the undesirable effect on the result of this study. If the processes that need to be applied equally to all samples, such as finishing and polishing, were carried out by different technicians, it would be difficult to ensure accurate comparisons among the groups since this study was based on an experiment requiring group comparison under the same conditions. Lastly, the manual designation of the border lines of the measurement areas can affect the measurement values finely. Owing to the nature of digital measurement in the metrology program, the value changes finely each time it is measured depending on the selected location. Therefore, we conducted measurements multiple times with careful selection of the border lines.

However, to date, metal 3D printing using SLM technology is the most convenient method for manufacturing RPD frameworks, which can reduce the time and labor required for conventional laboratory processes. To produce more accurate RPD frameworks than conventional cast frameworks, further studies using various metal 3D printers and software programs are essential. Further, more in vivo studies on the accuracy, fitness, and longevity of metal 3D-printed RPD frameworks are needed until the use of metal 3D-printed RPD frameworks becomes generalized.

Overall, the following conclusions were drawn in this study: (1) SLM-fabricated RPD frameworks exhibited an overall accuracy similar to that of conventional cast RPD frameworks; and (2) a combined method of resin 3D printing and casting demonstrated inferior accuracy. However, all frameworks in the three groups were clinically acceptable.

Supplementary Materials: The following supporting information can be downloaded at https://www.mdpi.com/article/10.3390/ma17133148/s1, Table S1: The internal discrepancies (μm) at three rests and four tissue stops in mandibular RPD metal frameworks of three groups; Figure S1: Comparisons of the internal discrepancies at the rests of #35, #44, #47 in mandibular RPD metal frameworks fabricated using three methods (SLM: selective laser melting-based metal 3D printing, RPC: DLP-based resin 3D printing and subsequent casting, CON: conventional lost-wax casting); Figure S2: Comparisons of the internal discrepancies at the tissue stops of #36, #37, #45, #46 in mandibular RPD metal frameworks fabricated using three methods (SLM: selective laser melting-based metal 3D printing, RPC: DLP-based resin 3D printing and subsequent casting, CON: conventional lost-wax casting).

Author Contributions: Conceptualization, J.-H.K.; Data curation, S.K.; Formal analysis, K.C.O.; Investigation, S.K.; Methodology, K.C.O.; Project administration, J.-H.K.; Resources, S.K.; Supervision, K.C.O. and J.-H.K.; Validation, J.-H.K.; Visualization, S.K.; Writing—original draft, S.K. and K.C.O.; Writing—review and editing, J.-H.K. All authors have read and agreed to the published version of the manuscript.

Funding: This research was funded by the National Research Foundation of Korea (NRF) grant funded by the Korea government (MSIT) (No. 2021R1F1A1055901) and by the Basic Science Research Program through the NRF funded by the Ministry of Education (No. 2021R1I1A1A01048233).

Institutional Review Board Statement: Not applicable.

Informed Consent Statement: Not applicable.

Data Availability Statement: The data presented in this study are available on request from the corresponding author.

Conflicts of Interest: The authors declare no conflicts of interest.

References

1. Girardot, R.L. History and Development of Partial Denture Design. *J. Am. Dent. Assoc.* **1941**, *28*, 1399–1408. [CrossRef]
2. Becker, C.M.; Kaiser, D.A.; Goldfogel, M.H. Evolution of removable partial denture design. *J. Prosthodont.* **1994**, *3*, 158–166. [CrossRef] [PubMed]
3. Ucar, Y.; Akova, T.; Akyil, M.S.; Brantley, W.A. Internal fit evaluation of crowns prepared using a new dental crown fabrication technique: Laser-sintered Co-Cr crowns. *J. Prosthet. Dent.* **2009**, *102*, 253–259. [CrossRef] [PubMed]
4. van Noort, R. The future of dental devices is digital. *Dent. Mater.* **2012**, *28*, 3–12. [CrossRef]
5. Venkatesh, K.V.; Nandini, V.V. Direct metal laser sintering: A digitised metal casting technology. *J. Indian Prosthodont. Soc.* **2013**, *13*, 389–392. [CrossRef] [PubMed]

6. Diwan, R.; Talic, Y.; Omar, N.; Sadig, W. Pattern waxes and inaccuracies in fixed and removable partial denture castings. *J. Prosthet. Dent.* **1997**, *77*, 553–555. [CrossRef] [PubMed]
7. Rekow, D. Computer-aided design and manufacturing in dentistry: A review of the state of the art. *J. Prosthet. Dent.* **1987**, *58*, 512–516. [CrossRef] [PubMed]
8. Dawood, A.; Marti Marti, B.; Sauret-Jackson, V.; Darwood, A. 3D printing in dentistry. *Br. Dent. J.* **2015**, *219*, 521–529. [CrossRef]
9. Azari, A.; Nikzad, S. The evolution of rapid prototyping in dentistry: A review. *Rapid Prototyp. J.* **2009**, *15*, 216–225. [CrossRef]
10. Lima, J.M.; Anami, L.C.; Araujo, R.M.; Pavanelli, C.A. Removable partial dentures: Use of rapid prototyping. *J. Prosthodont.* **2014**, *23*, 588–591. [CrossRef]
11. Beguma, Z.; Chhedat, P. Rapid prototyping--when virtual meets reality. *Int. J. Comput. Dent.* **2014**, *17*, 297–306. [PubMed]
12. Bae, E.J.; Jeong, I.D.; Kim, W.C.; Kim, J.H. A comparative study of additive and subtractive manufacturing for dental restorations. *J. Prosthet. Dent.* **2017**, *118*, 187–193. [CrossRef] [PubMed]
13. Suzuki, Y.; Shimizu, S.; Waki, T.; Shimpo, H.; Ohkubo, C. Laboratory efficiency of additive manufacturing for removable denture frameworks: A literature-based review. *Dent. Mater. J.* **2021**, *40*, 265–271. [CrossRef] [PubMed]
14. Koutsoukis, T.; Zinelis, S.; Eliades, G.; Al-Wazzan, K.; Rifaiy, M.A.; Al Jabbari, Y.S. Selective Laser Melting Technique of Co-Cr Dental Alloys: A Review of Structure and Properties and Comparative Analysis with Other Available Techniques. *J. Prosthodont.* **2015**, *24*, 303–312. [CrossRef] [PubMed]
15. Barazanchi, A.; Li, K.C.; Al-Amleh, B.; Lyons, K.; Waddell, J.N. Additive Technology: Update on Current Materials and Applications in Dentistry. *J. Prosthodont.* **2017**, *26*, 156–163. [CrossRef] [PubMed]
16. Tregerman, I.; Renne, W.; Kelly, A.; Wilson, D. Evaluation of removable partial denture frameworks fabricated using 3 different techniques. *J. Prosthet. Dent.* **2019**, *122*, 390–395. [CrossRef] [PubMed]
17. Revilla-Leon, M.; Ozcan, M. Additive Manufacturing Technologies Used for Processing Polymers: Current Status and Potential Application in Prosthetic Dentistry. *J. Prosthodont.* **2019**, *28*, 146–158. [CrossRef]
18. Snosi, A.M.; Lotfy, S.M.; Thabet, Y.G.; Sabet, M.E.; Rizk, F.N. Subtractive versus additive indirect manufacturing techniques of digitally designed partial dentures. *J. Adv. Prosthodont.* **2021**, *13*, 327–332. [CrossRef] [PubMed]
19. Williams, R.J.; Bibb, R.; Rafik, T. A technique for fabricating patterns for removable partial denture frameworks using digitized casts and electronic surveying. *J. Prosthet. Dent.* **2004**, *91*, 85–88. [CrossRef]
20. Bibb, R.; Eggbeer, D.; Williams, R. Rapid manufacture of removable partial denture frameworks. *Rapid Prototyp. J.* **2006**, *12*, 95–99. [CrossRef]
21. Williams, R.J.; Bibb, R.; Eggbeer, D.; Collis, J. Use of CAD/CAM technology to fabricate a removable partial denture framework. *J. Prosthet. Dent.* **2006**, *96*, 96–99. [CrossRef] [PubMed]
22. Alageel, O.; Abdallah, M.N.; Alsheghri, A.; Song, J.; Caron, E.; Tamimi, F. Removable partial denture alloys processed by laser-sintering technique. *J. Biomed. Mater. Res. B Appl. Biomater.* **2018**, *106*, 1174–1185. [CrossRef] [PubMed]
23. Stamenkovic, D.; Popovic, M.; Rudolf, R.; Zrilic, M.; Raic, K.; Duricic, K.O.; Stamenkovic, D. Comparative Study of the Microstructure and Properties of Cast-Fabricated and 3D-Printed Laser-Sintered Co-Cr Alloys for Removable Partial Denture Frameworks. *Materials* **2023**, *16*, 3267. [CrossRef] [PubMed]
24. Takaichi, A.; Fueki, K.; Murakami, N.; Ueno, T.; Inamochi, Y.; Wada, J.; Arai, Y.; Wakabayashi, N. A systematic review of digital removable partial dentures. Part II: CAD/CAM framework, artificial teeth, and denture base. *J. Prosthodont. Res.* **2022**, *66*, 53–67. [CrossRef] [PubMed]
25. Zhou, Y.; Li, N.; Yan, J.; Zeng, Q. Comparative analysis of the microstructures and mechanical properties of Co-Cr dental alloys fabricated by different methods. *J. Prosthet. Dent.* **2018**, *120*, 617–623. [CrossRef] [PubMed]
26. Lapcevic, A.R.; Jevremovic, D.P.; Puskar, T.M.; Williams, R.J.; Eggbeer, D. Comparative analysis of structure and hardness of cast and direct metal laser sintering produced Co-Cr alloys used for dental devices. *Rapid Prototyp. J.* **2016**, *22*, 144–151. [CrossRef]
27. Souza Curinga, M.R.; Claudino Ribeiro, A.K.; de Moraes, S.L.D.; do Egito Vasconcelos, B.C.; da Fonte Porto Carreiro, A.; Pellizzer, E.P. Mechanical properties and accuracy of removable partial denture frameworks fabricated by digital and conventional techniques: A systematic review. *J. Prosthet. Dent.* **2023**. [CrossRef] [PubMed]
28. Carneiro Pereira, A.L.; Bezerra de Medeiros, A.K.; de Sousa Santos, K.; Oliveira de Almeida, E.; Seabra Barbosa, G.A.; da Fonte Porto Carreiro, A. Accuracy of CAD-CAM systems for removable partial denture framework fabrication: A systematic review. *J. Prosthet. Dent.* **2021**, *125*, 241–248. [CrossRef] [PubMed]
29. Ahmed, N.; Abbasi, M.S.; Haider, S.; Ahmed, N.; Habib, S.R.; Altamash, S.; Zafar, M.S.; Alam, M.K. Fit Accuracy of Removable Partial Denture Frameworks Fabricated with CAD/CAM, Rapid Prototyping, and Conventional Techniques: A Systematic Review. *Biomed. Res. Int.* **2021**, *2021*, 3194433. [CrossRef]
30. Soltanzadeh, P.; Suprono, M.S.; Kattadiyil, M.T.; Goodacre, C.; Gregorius, W. An In Vitro Investigation of Accuracy and Fit of Conventional and CAD/CAM Removable Partial Denture Frameworks. *J. Prosthodont.* **2019**, *28*, 547–555. [CrossRef]
31. Ye, H.; Ning, J.; Li, M.; Niu, L.; Yang, J.; Sun, Y.; Zhou, Y. Preliminary Clinical Application of Removable Partial Denture Frameworks Fabricated Using Computer-Aided Design and Rapid Prototyping Techniques. *Int. J. Prosthodont.* **2017**, *30*, 348–353. [CrossRef] [PubMed]
32. Arnold, C.; Hey, J.; Schweyen, R.; Setz, J.M. Accuracy of CAD-CAM-fabricated removable partial dentures. *J. Prosthet. Dent.* **2018**, *119*, 586–592. [CrossRef] [PubMed]

33. Negm, E.E.; Aboutaleb, F.A.; Alam-Eldein, A.M. Virtual Evaluation of the Accuracy of Fit and Trueness in Maxillary Poly(etheretherketone) Removable Partial Denture Frameworks Fabricated by Direct and Indirect CAD/CAM Techniques. *J. Prosthodont.* **2019**, *28*, 804–810. [CrossRef]
34. Lee, J.W.; Park, J.M.; Park, E.J.; Heo, S.J.; Koak, J.Y.; Kim, S.K. Accuracy of a digital removable partial denture fabricated by casting a rapid prototyped pattern: A clinical study. *J. Prosthet. Dent.* **2017**, *118*, 468–474. [CrossRef]
35. Oh, K.C.; Yun, B.S.; Kim, J.H. Accuracy of metal 3D printed frameworks for removable partial dentures evaluated by digital superimposition. *Dent. Mater.* **2022**, *38*, 309–317. [CrossRef]
36. Tasaka, A.; Shimizu, T.; Kato, Y.; Okano, H.; Ida, Y.; Higuchi, S.; Yamashita, S. Accuracy of removable partial denture framework fabricated by casting with a 3D printed pattern and selective laser sintering. *J. Prosthodont. Res.* **2020**, *64*, 224–230. [CrossRef]
37. Bajunaid, S.O.; Altwaim, B.; Alhassan, M.; Alammari, R. The Fit Accuracy of Removable Partial Denture Metal Frameworks Using Conventional and 3D Printed Techniques: An In Vitro Study. *J. Contemp. Dent. Pract.* **2019**, *20*, 476–481. [CrossRef]
38. Chen, H.; Li, H.; Zhao, Y.; Zhang, X.; Wang, Y.; Lyu, P. Adaptation of removable partial denture frameworks fabricated by selective laser melting. *J. Prosthet. Dent.* **2019**, *122*, 316–324. [CrossRef]
39. Almufleh, B.; Emami, E.; Alageel, O.; de Melo, F.; Seng, F.; Caron, E.; Nader, S.A.; Al-Hashedi, A.; Albuquerque, R.; Feine, J.; et al. Patient satisfaction with laser-sintered removable partial dentures: A crossover pilot clinical trial. *J. Prosthet. Dent.* **2018**, *119*, 560–567.e561. [CrossRef] [PubMed]
40. Peng, P.W.; Hsu, C.Y.; Huang, H.Y.; Chao, J.C.; Lee, W.F. Trueness of removable partial denture frameworks additively manufactured with selective laser melting. *J. Prosthet. Dent.* **2022**, *127*, 122–127. [CrossRef]
41. Stern, M.A.; Brudvik, J.S.; Frank, R.P. Clinical evaluation of removable partial denture rest seat adaptation. *J. Prosthet. Dent.* **1985**, *53*, 658–662. [CrossRef] [PubMed]
42. Dunham, D.; Brudvik, J.S.; Morris, W.J.; Plummer, K.D.; Cameron, S.M. A clinical investigation of the fit of removable partial dental prosthesis clasp assemblies. *J. Prosthet. Dent.* **2006**, *95*, 323–326. [CrossRef] [PubMed]

materials

Article

Metal 3D-Printed Bioinspired Lattice Elevator Braking Pads for Enhanced Dynamic Friction Performance

Nikolaos Kladovasilakis [1,*], Eleftheria Maria Pechlivani [1], Ioanna K. Sfampa [2], Konstantinos Tsongas [3], Apostolos Korlos [3], Constantine David [4] and Dimitrios Tzovaras [1]

1 Centre for Research and Technology Hellas, Information Technologies Institute (CERTH/ITI), 57001 Thessaloniki, Greece; riapechl@iti.gr (E.M.P.); dimitrios.tzovaras@iti.gr (D.T.)
2 R&D Department, KLEEMANN Group, 61100 Kilkis, Greece; i.sfampa@kleemannlifts.com
3 Advanced Materials and Manufacturing Technologies Laboratory, Department Industrial Engineering and Management, School of Engineering, International Hellenic University, 57001 Thessaloniki, Greece; ktsongas@ihu.gr (K.T.); apkorlos@ihu.gr (A.K.)
4 Manufacturing Technology and Production Systems Laboratory, Department of Mechanical Engineering, International Hellenic University, 62124 Serres, Greece; david@ihu.gr
* Correspondence: nikoklad@iti.gr

Abstract: The elevator industry is constantly expanding creating an increased demand for the integration of high technological tools to increase elevator efficiency and safety. Towards this direction, Additive Manufacturing (AM), and especially metal AM, is one of the technologies that could offer numerous competitive advantages in the production of industrial parts, such as integration of complex geometry, high manufacturability of high-strength metal alloys, etc. In this context, the present study has 3D designed, 3D printing manufactured, and evaluated novel bioinspired structures for elevator safety gear friction pads with the aim of enhancing their dynamic friction performance and eliminating the undesired behavior properties observed in conventional pads. Four different friction pads with embedded bioinspired surface lattice structures were formed on the template of the friction surface of the conventional pads and 3D printed by the Selective Laser Melting (SLM) process utilizing tool steel H13 powder as feedstock material. Each safety gear friction pad underwent tribological tests to evaluate its dynamic coefficient of friction (CoF). The results indicated that pads with a high contact surface area, such as those with car-tire-like and extended honeycomb structures, exhibit high CoF of 0.549 and 0.459, respectively. Based on the acquired CoFs, Finite Element Models (FEM) were developed to access the performance of braking pads under realistic operation conditions, highlighting the lower stress concentration for the aforementioned designs. The 3D-printed safety gear friction pads were assembled in an existing emergency progressive safety gear system of KLEEMANN Group, providing sufficient functionality.

Keywords: bioinspired surface lattices; additive manufacturing; selective laser melting; dynamic friction; elevator's braking pads; tribological analysis; finite element models

Citation: Kladovasilakis, N.; Pechlivani, E.M.; Sfampa, I.K.; Tsongas, K.; Korlos, A.; David, C.; Tzovaras, D. Metal 3D-Printed Bioinspired Lattice Elevator Braking Pads for Enhanced Dynamic Friction Performance. *Materials* **2024**, *17*, 2765. https://doi.org/10.3390/ma17112765

Academic Editor: Yong-Cheng Lin

Received: 17 May 2024
Revised: 29 May 2024
Accepted: 4 June 2024
Published: 5 June 2024

1. Introduction

The elevator industry is constantly expanding, driven by the development of cities and the increasing demand for high-rise buildings [1,2]. Concurrently, the imperative for safer and more efficient vertical transportation in buildings of all types persists. Technological advancements in this sector enable the development of elevators with superior performance and safety standards. Adherence to safety regulations is paramount in both the construction and operation phases of elevators. Mechanical components, such as emergency braking systems and safety gears, play a critical role in elevator safety and are required by national and European regulations [3,4].

Progressive safety gear is a mechanical device used to deal with situations of potential danger, such as exceeding the descent speed or breaking the means of suspension, for example, wire ropes [5,6]. In these cases, an innovative grab is activated automatically, ensuring the immediate stop of the elevator cabin movement and protecting the occupants from possible accidents. In detail, this progressive safety gear can be placed both under the cabin, to avoid sudden downward and/or upward movement, and under the counterweight, in the event that the lower space in the shaft, and in particular under the travel of the counterweight, is accessible. The basic elements of the progressive safety gear include the friction plate (pad), the housing, disc springs, and the gripper wheel. In order to avoid over-acceleration of the elevator, especially for elevators with a nominal speed of more than 1 m/s, the progressive safety gear is used according to the state of the art. Figure 1 shows an indicative image of an existing progressive safety gear system along with its position on the elevator's system. Specifically, the following phases can be distinguished in the operation of the progressive safety gear:

i. The resting phase, where the wheel is at a safe distance (a few millimeters) from the guide rails, while the friction plate is at a short distance of 1–3 mm from the guide rails.
ii. Activation phase, where the wheel has just touched the guide rails
iii. The wedging phase is where the wheel moves and comes into contact with the guide rails until the gap between the guide rails and the back is zero, at which point the safety gear begins to brake.
iv. After the guide rails are in contact with the friction plate, the springs are compressed and thus the friction forces develop gradually, resulting in progressive braking.

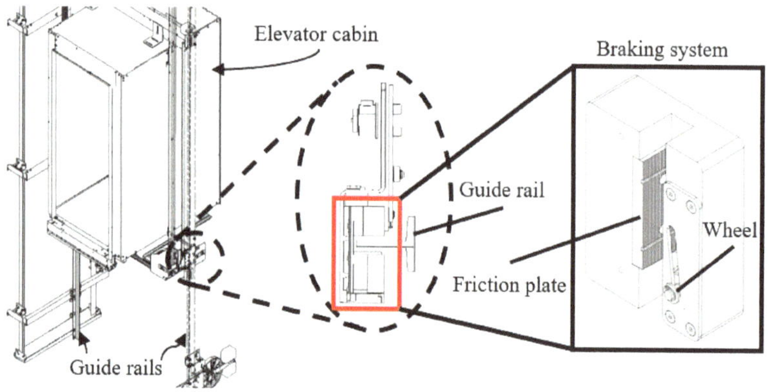

Figure 1. Image of an indicative existing progressive emergency braking system along with its position on the elevator's system.

The main disadvantages of existing commercial safety gears in the elevator progressive emergency braking system are the damages caused both on the safety gear itself and the guide rails during activation. More specifically, when the safety gear of an elevator is activated, the friction plates strongly deform the guide rails' surfaces leading to complete failure after one or at most two times of emergency braking, forcing their replacement. These incidents increase the cost of maintenance and the operational costs of the elevator due to the downtime period, which in many cases is highly undesirable, especially in buildings where the elevator system is a vital infrastructure, such as in hospitals. In addition, with the existing morphology of the commercial friction plates, it becomes difficult to predict the behavior of the material, since instead of achieving immobilization of the cabin (or counterweight) through the frictional forces developed; the results often are the plastic deformation of the guide rails due to partial penetration of the friction plate into guide rails body. This behavior also leads to an undesirable experience for the passengers of the elevator, increasing the possibility of their injury. In order to address the aforementioned issues, in recent decades, both the research community and the elevator industry examined and developed novel solutions optimizing the performance and reliability of the progressive safety gear. There are numerous submitted patents in this direction. However, there is a lack of published literature about the examination of innovative patterns in the contact area of a friction plate to increase its dynamic frictional performance.

In this context, the current study proposes the technical solution of integrating hierarchical biomimetic patterns on the surface of a friction pad in order to develop and manufacture pioneering friction pads, with optimal production techniques and advanced materials. The employment of this novel approach is expected to provide a smoother immobilization of the elevator, reducing the possibility of injury of the passengers, and avoiding the unpleasant experience of a sudden and emergency elevator stop. In addition, it is anticipated to increase the lifetime of the specific components, due to the existence of large diffusion canals, reducing the maintenance intervals and costs of an elevator. More specifically, in this paper, the development of four innovative friction plates for the safety gear of the elevator emergency braking system were developed. These advanced morphologies were based on the application of bio-inspired hierarchical patterns, i.e., specific repeated geometric shapes, which can significantly improve the tribological behavior of a surface. This innovation of hierarchical patterns design can be incorporated into various types of elevator safety gear offering comparative advantages. In detail, the superiority of these new morphologies of the present study is in the application of a biomimetic friction surface design process, i.e., in copying the morphology of natural patterns with proven high adhesion, such as the sole of a frog's rear ends. The developed morphologies are the honeycomb, extended honeycomb, speckled honeycomb, and car-tire geometries with a biomimetic design. The designed friction pads were additively manufactured using the Selective Laser Melting (SLM) technique and a tribological analysis was performed for each one. Then, based on the results of tribological analyses, i.e., coefficient of friction (CoF), the produced braking pads were numerically evaluated via Finite Element Analyses (FEAs) under realistic loads. Figure 2 presents the flowchart of the current study.

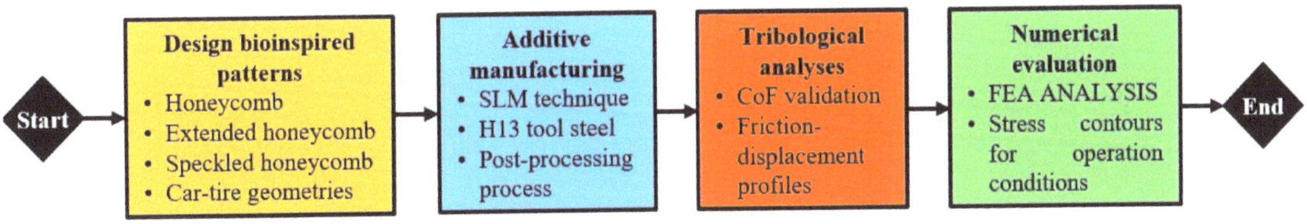

Figure 2. Flowchart of the current study.

2. Materials and Methods

2.1. Problem Statement

Progressive safety gears are widely utilized in the elevator industry, increasing the safety and reliability of elevator systems. Typically, the progressive safety gear of a commercial elevator consists of two friction pads with a trapezoidal configuration to achieve progressive deceleration. In this way, the immobilization of the elevator occurs smoothly avoiding potential injuries to the passengers. However, in order to achieve the immobilization of an elevator, there is a need to create a friction force (opposite to the movement of the elevator) on the guide rails greater than its overall weight force. According to existing standardization [7], the overall weight force of an elevator is provided by the following equation (Equation (1)). Where P is the mass of the empty cabin which includes the masses of doors, walls, floor, roof, and cabin sling. On the other hand, Q is the payload of an elevator, which is the maximum weight it can carry for the selected cabin [8]. Furthermore, g is the gravitational acceleration (9.81 m/s^2) and FOS is the factor of safety which ensures that deceleration of the elevator occurs. The value of this property is usually around 2.

$$Total\ Braking\ Force = (P + Q) \times g \times FOS \quad (1)$$

Therefore, the braking force (BF) for each safety gear is evaluated by the following equation. This equation indicates the necessary vertically applied force (VF) from the friction pads to guide rails in order to achieve the desired friction force (Equation (3)). Where μ is the experimental coefficient of friction between the friction pad and the guide rail.

$$BF = \frac{Total\ Braking\ Force}{2} \tag{2}$$

$$VF = \frac{BF}{\mu} \tag{3}$$

From the aforementioned analysis, it is obvious that in order to immobilize an elevator, a sufficient amount of vertical force should be applied on the guide rails, much higher than the weight force, due to the fact that the coefficient of friction ranges between 0 and 1. This amount of vertical force often severely damages the guide rails, increasing dramatically the maintenance cost after an emergency brake activation. Therefore, elevator manufacturers explore novel ways to reduce the amount of vertical force. The most straightforward technique to reduce the vertical force is to increase the coefficient of friction between the friction pads and the guide rails. According to the datasheet of the KLEEMANN Group elevator manufacturer, the existing friction pads exhibit a coefficient of friction of around 0.3. Hence, in the scope of this paper, novel surface architected materials are employed on the surface of the friction pad in order to increase the coefficient of friction and thereby reduce the applied vertical force offering efficient and smooth elevator immobilization without damaging its guiding system.

2.2. Design of Bio-Inspired Braking Pads

In the context of this study, four different surface-architected materials were employed with the aim of increasing the surface's coefficient of friction. These structures were inspired by natural structures with high dynamic friction performance, such as the pads of a frog's back legs and the pads of cheetah legs for honeycomb structures and for car-tire-like structures, respectively [9,10]. The selected surface-architected materials, i.e., extended honeycombs, honeycombs, speckled honeycombs, and a car-tire-like structure, were integrated into the structure of existing friction pads as it is presented in Figure 3. In general, honeycomb configurations are employed in the automotive industry [11] for braking applications showing enhanced braking performance, therefore, three of the four chosen structures belong to this category. On the other hand, the car-tire structure has been designed to have traction on the road, thus the corresponding structure (car-tire-like) was a suitable candidate for an elevator braking pad. The concept behind the selection of these structures was to create consecutive extended contact surfaces with sufficient channels for debris removal. In this way, the accumulated friction increases the coefficient of friction between the pad and the smooth surface of the guide rail minimizing the damage due to the reduced vertical force and the uniform removal of metal filings during the braking process. The designed dimensions of the employed structures were chosen in order to offer high printability for the developed friction pads and ensure sufficient diffusion channel width. The thickness of each structure's elements (i.e., for unit cell) was 2 mm in order to provide an extended life cycle (i.e., more than one use) for the braking pad without compromising their mechanical performance. It is worth mentioning that the design process was performed in the SolidWorks™ design platform (Dassault Systèmes SE, Vélizy-Villacoublay, France).

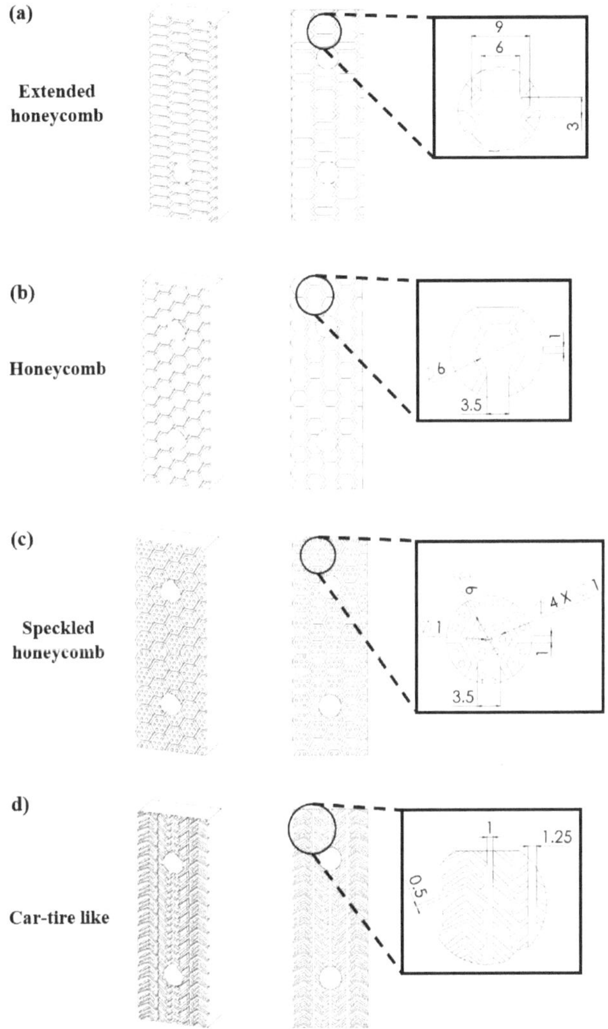

Figure 3. Isometric, front, and detailed view of the designed surface architected materials of (unit: mm): (**a**) Extended honeycomb; (**b**) Honeycomb; (**c**) Speckled honeycomb; (**d**) Car-tire-like configurations.

2.3. Material Selection and Additive Manufacturing

After the design of friction pads, the next stage is the manufacturing process for these parts. The first step in this process is the material selection. The existing friction pads are fabricated from tool steels in order to withstand the increased braking loads. In this context, for the developed friction pads, the high-strength H13 tool steel was selected as the construction material. The upper section of Table 1 lists the basic properties of the H13 tool steel. These properties were provided by the manufacturer (OC Oerlikon, Freienbach, Switzerland) of the raw material for metal additive manufacturing and from a previously published study on 3D-printed H13 tool steel [12].

Table 1. PSD, Hall Flow, and Chemical Composition for H13 tool steel powder according to manufacturer.

Properties			Values		
Density			7.80 g/cm^3		
Elastic modulus			200,000 ± 5000 MPa		
Poisson ratio			0.29		
Yield Strength (ASTM E8) [13]			1538 ± 31 MPa		
Ultimate Tensile Strength (ASTM E8) [13]			1903 ± 12 MPa		
Elongation at break (ASTM E8) [13]			11 ± 1 %		
Nominal Range (μm)	**D90 (μm)**		**D50 (μm)**	**D10 (μm)**	**Hall Flow (s/50 g)**
−45 + 15	50		34	21	≤25
Weight Percent (nominal)					
Fe	Cr	Mo	Si	V	C
Balance	5.2	1.3	1.0	1.0	0.4
3D Printing Parameters			**Values**		
Layer height			0.25 μm		
Hatching distance			0.40 μm		
Laser power			150 W		
Print speed			1000 mm/s		
VED			150 J/mm^3		
Beam diameter			0.40 μm		

Therefore, for the fabrication of the developed braking pads, the SLM additive manufacturing technique was employed utilizing the ORLAS Creator (Coherent Inc., Santa Clara, CA, USA) metal 3D printer. The ORLAS Creator is equipped with a continuous 250 W Yb-fiber laser beam with a wavelength of 1067 nm to sufficiently melt the metal powder. Moreover, its maximum printing accuracy reaches up to 25 μm at the vertical direction/z axis (layer height) and 40 μm in the XY plane (hatching distance) [14]. The selected material was in the form of fine powder and its basic characteristics of particle size distribution (PSD), Hall flow, and chemical composition are listed in Table 1. This combination of parameters ensured the high printability and quality of the 3D-printed friction pads. Furthermore, in order to achieve the optimum quality in terms of mechanical behavior and accuracy, specific process-related parameters were selected based on the results of previous studies (Bottom side of Table 1). It is worth noting that Equation (4) describes the mathematical relation between one of the most important parameters for the fusion of the powder, i.e., the volumetric energy density (*VED*), with the basic process-related parameters, such as layer height (*l*), hatching distance (*h*), laser power (*P*), and printing speed (*V*). Based on the value of VED, we ensured the proper thermal energy exposure for the powder in order to create sufficient and uniform molten pools, which increase the relative density of the part (>99.8%) and eliminate the potential defects (un-melted particles, keyholes, voids, etc.) [15]. Finally, printing orientation was chosen to be ZXY, according to international standards [16], in order for the printed parts to present the maximum strength under operation conditions.

$$VED = \frac{P}{V \cdot h \cdot l} \tag{4}$$

2.4. Tribological Analysis

The first step before conducting the tribological analysis was the roughness measurements of the surface of the 3D-printed parts. The surface roughness analysis was performed using the Wyko® NT Series optical profiler (Veeco Instruments Inc., Plainview, NY, USA), NT1100 Optical Profiling System. This system employs white light interferometry for high-resolution 3D surface roughness measurements. Specimens' profiles were assessed utilizing the optical surface profilometer, specifically focusing on the friction surface of the 3D-printed H13 pads. Discussing surface roughness alongside tribological results is important because it directly influences friction and wear behavior, providing crucial insights into material interactions and performance. It is worth noting that the most crucial roughness parameters for the specimen's surface and profile were obtained according to international standards [17,18].

In a series of tribological experiments conducted by a Tribolab Universal Material Testing (UMT) apparatus, employing a pin diameter of 6.56 mm, a standardized normal force of 50 N was applied to a distance of 10 mm with a sliding velocity of 1 mm/s. These meticulously controlled parameters served as the foundation for a rigorous examination of frictional dynamics and wear characteristics inherent to diverse material compositions [19,20]. Utilizing the advanced capabilities of the UMT TriboLab from Bruker (Billerica, MA, USA), a pin-on-disc configuration was employed to investigate the tribological behavior of AM H13 steel specimens. These specimens were designed as safety braking pads, featuring various configurations including honeycomb, honeycomb speckled, extended honeycomb, car-tire-like, and pyramid structures. The focus of the investigation centered on determining the CoF, a widely accepted standard parameter of friction. This parameter is calculated from the force sensor signal, considering both the normal and longitudinal forces acquired simultaneously. By analyzing the CoF, insights into the frictional characteristics of each configuration were acquired, facilitating a comprehensive understanding of their tribological performance. This setup facilitated a focused inquiry into the tribological properties of these design patterns examined, aiming to discern their performance under controlled conditions. All tests were performed at least five times at room temperature, on the region between fixation holes.

2.5. Finite Element Models

In order to evaluate fast, accurately, and reliably the performance of the developed friction pads, finite element models (FEMs) were developed based on the results of tribological analysis. For this purpose, the ANSYS Workbench (ANSYS, Inc., Canonsburg, PA, USA) simulation platform was utilized and, more specifically, its static module. The employed material model was a bilinear isotropic hardening in which the elastic section emerged from the aforementioned material properties and the plastic section occurred after the yield stress point with a tangent modulus of 10,000 MPa based on the experimental results of a previous study [12]. The next step was the creation of the computational mesh, which was defined after the conduction of mesh sensitivity analyses regarding the maximum von Mises stress in order to extract mesh-independent results. This analysis led to mesh with tetrahedral elements and a minimum size of 0.4 mm with total mesh elements ranging from around 100,000 to 125,000 elements depending on the applied geometry of the structure. Regarding the fixation condition of the analyses, each friction pad was set to have zero displacements on the plane of each support hole and zero displacements on the back of each pad where the spring set was applied. On the other hand, for the loading condition, the worst-case scenario of a commercial passenger elevator was chosen with P + Q at 2600 kg and two progressive safety gears. Moreover, the overall force applied to the friction surface of each pad was analyzed in two different components, namely the vertical force and braking/friction force. The braking/friction force has a direction opposite to the movement of the elevator and it is constant due to the fact that the same value for this force results in the immobilization of the elevator regardless of the applied surface pattern of each pad. In contrast, the necessary vertical force in order to create the friction force is dependent on the coefficient of friction of each developed friction pad. Therefore, the evaluation of this force component for each case was performed utilizing the results of the tribological analysis. It is worth mentioning that in the context of this study, a commercial friction pad was also tested in order to provide a comprehensive benchmark for the designed braking pads. Figure 4 presents an indicative example of the commercial friction pad with the fixation and loading conditions that have been considered in the developed FEM.

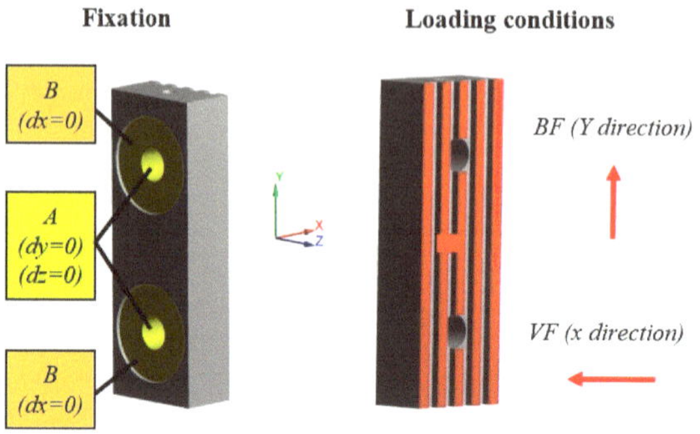

Figure 4. Fixation and loading conditions on the commercial friction pad for the developed FEMs.

3. Results

3.1. 3D-Printed Friction Pads

The developed friction pads with advanced surface geometry were additively manufactured via the SLM 3D printing technique, as shown in Figure 5a. The 3D printing process was performed in a low concentration of oxygen (<1%) to avoid oxidation phenomena. Moreover, all four designed friction pads were fitted on the 3D printer's platform in ZYX orientation as presented in Figure 5b. The overall 3D printing process lasted around 13 h indicating the rapid fabrication nature of AM technology. It is worth mentioning that the 3D prints were structurally flawless, without any visible defects and voids. Furthermore, the 3D-printed specimens were weighted with a high-accuracy scale and their weight values were found close to nominal values revealing negligible porosity for the structures. Then, the 3D-printed friction pads underwent the post-processing treatment of support removal and sandblasting in order to be finished. Finally, the developed friction pads were assembled in the system of a safety gear showing sufficient fit and functionality, as illustrated in Figure 5c.

Figure 5. (**a**) The developed 3D-printed friction pads; (**b**) The friction pads inside the building chamber; (**c**) Indicative images of the assembly between the developed friction pads and the elevator safety gear.

3.2. Tribological Results

Before the tribological analysis occurs, the roughness of the 3D-printed friction pads needs to be evaluated as it is strongly related to the CoF. The roughness measurements were performed on the contact surface of each specimen. Figure 6 illustrates an indicative example of the roughness evaluation on the 3D-printed parts. In detail, with the aid of the optical profilometer, both the surface roughness contour and roughness profiles were obtained. These acquired data show that the 3D-printed components in the examined plane (i.e., XZ plane) have roughness fluctuating between 0 and 15 μm with smooth transitions.

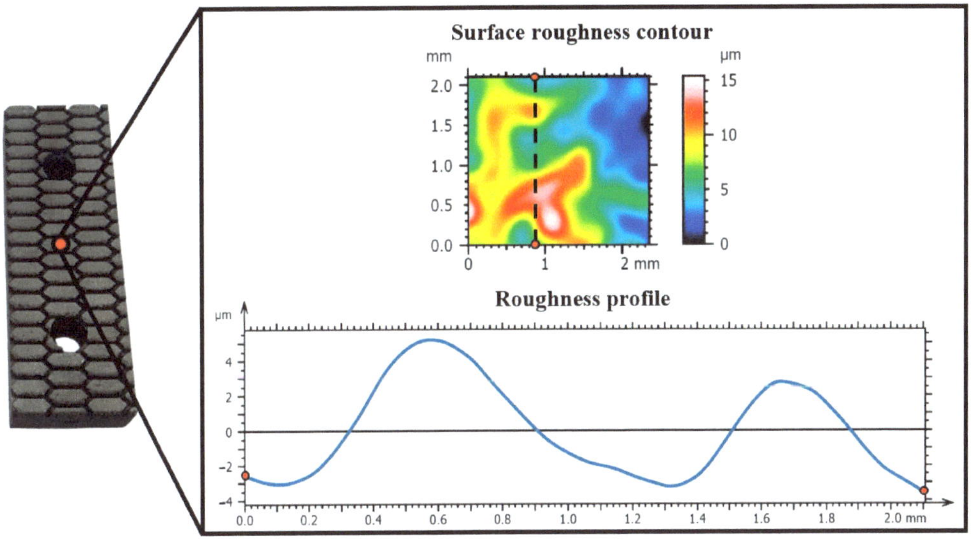

Figure 6. Surface roughness contour and roughness profile of an indicative braking surface of the 3D-printed pads.

It is worth noting that the examined area for roughness is independent of the patterns as it is an order of magnitude smaller (around 2 × 2 mm). Therefore, the 3D-printed surface was measured in roughness analysis which has similar value roughness for all developed braking pads due to the repeatability of the SLM 3D printing process. In addition, Table 2 lists all the crucial roughness parameters, i.e., average roughness (S_a, R_a), the root means square of the ordinate values (S_q, R_q), the mean roughness depth (S_z, R_z), the maximum pit height (S_v, R_v), and the maximum peak height (S_p, R_p).

Table 2. Mean values of roughness indicators for the 3D-printed braking pads.

Surface Roughness		Profile Roughness	
Indicators	**Values (μm)**	**Indicators**	**Values (μm)**
S_a	2.675 ± 0.1	R_a	0.7081 ± 0.05
S_q	3.168 ± 0.1	R_q	0.8189 ± 0.05
S_z	15.42 ± 0.4	R_z	2.619 ± 0.1
S_v	6.860 ± 0.2	R_v	1.113 ± 0.1
S_p	8.557 ± 0.25	R_p	1.506 ± 0.1

After the roughness analysis of the 3D-printed surface of the developed friction pads, the next step was the conduction of tribological analysis. All specimens were examined under the same conditions and indicative images from this process are illustrated in Figure 7 for the car-tire-like pad (a) and for the honeycomb pad (b). It is obvious that the direction of the probe is the same as the elevator movement direction in order to obtain relatable tribological results.

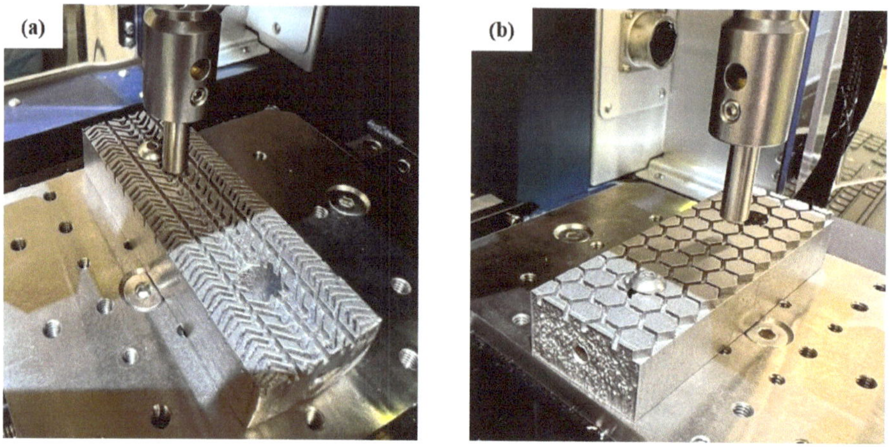

Figure 7. Indicative images of the (**a**) car-tire-like and (**b**) honeycomb friction pads during the pin-on-disc tribological tests.

Based on the tribological analysis, the honeycomb and honeycomb speckled configurations exhibited coefficients of friction (CoF) of 0.317 and 0.372, respectively. Remarkably, these values suggest a nearly identical pattern in their tribological behavior, owing to their closely matched CoF. However, the speckled variant displayed a slightly elevated CoF, likely attributable to the presence of speckles on its surface. This subtle distinction underscores the influence of surface texture on frictional dynamics, accentuating the importance of detailed analysis in tribological investigations. In the length of the sliding distance, we discern an excitation in Figure 8 due to the gap among the cells. For the extended honeycomb configuration, three excitations were observed, with the maximum CoF reaching 0.6. Additionally, the average CoF remained notably high, approximately around 0.459. In the car-tire-like configuration, the CoF is notably higher, reaching 0.549. This marked increase in CoF value highlighted the unique characteristics of the car-tire-like pattern, wherein its structural intricacies likely contributed to heightened frictional interaction. The car-tire-like pattern demonstrates the highest CoF, surpassing all the other configurations, solidifying its position as the configuration with the highest frictional propensity. This distinction underscores the diverse tribological behaviors manifested across different configurations, each bearing its own signature of frictional response.

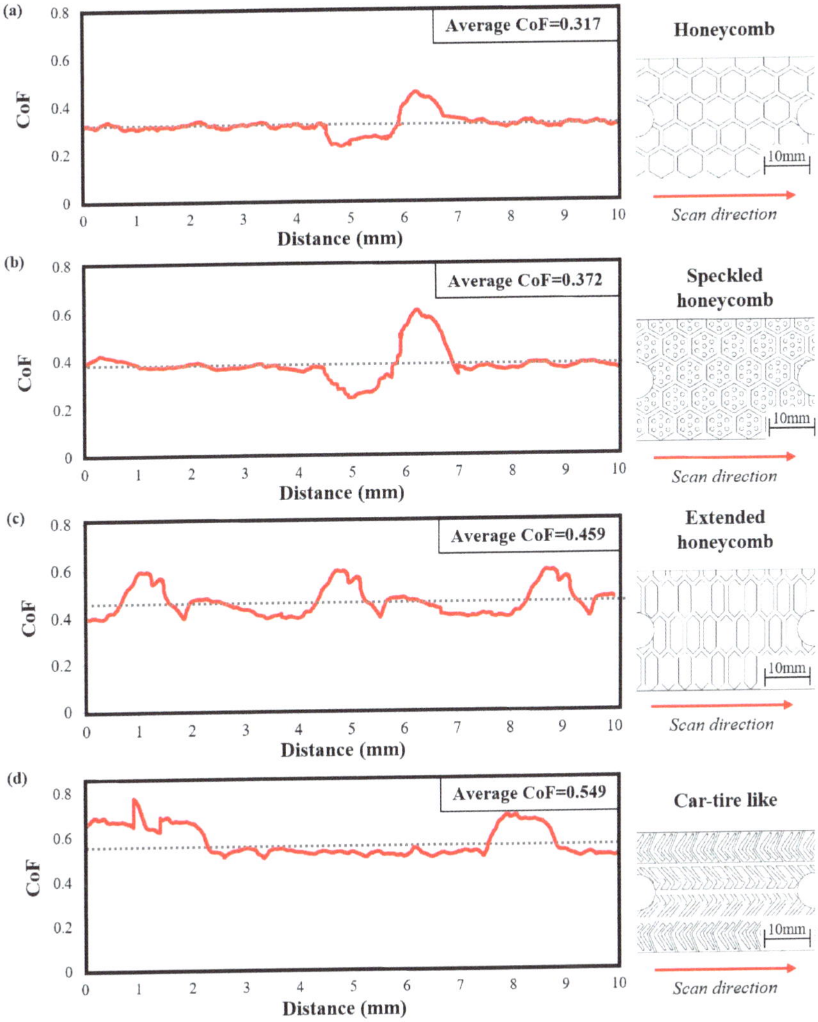

Figure 8. Tribological results of CoF to distance for pad's structures of: (**a**) Honeycomb; (**b**) Speckled honeycomb; (**c**) Extended honeycomb, and (**d**) Car-tire-like.

3.3. Finite Element Analyses

After the conduction of tribological analyses, the obtained results, notably the CoF, were used to accurately simulate the mechanical performance of each designed friction pad under realistic loads. Therefore, utilizing the measured CoF for each design, the corresponding vertical force was evaluated, as listed in Table 3. The evaluation process was performed based on the biggest standardized elevator system with $P + Q = 2600$ kg according to Equations (1)–(3). It is obvious that the necessary vertical force that should be applied in order to decelerate the elevator and eventually stop it is strongly related to the interface CoF. In particular, when the CoF at the interface is low, it necessitates an increase in vertical force exerted in the opposite direction stressing the braking pad.

Table 3. Vertical force evaluation based on CoF and necessary braking force for each structure.

Structure	Measured CoF	Braking Force (BF)	Vertical Force (VF)
Commercial	0.3		85,020 N
Honeycomb	0.317 ± 0.03		80,461 N
Speckled honeycomb	0.372 ± 0.04	25,506 N	68,565 N
Extended honeycomb	0.459 ± 0.05		55,569 N
Car-tire-like	0.549 ± 0.05		46,459 N

This observed behavior can easily result in the conclusion that the friction pads with low CoF are stressed more than the ones with higher CoF, leading to faster and abrupt part failure. Indeed, the conducted FEAs have shown similar results, as presented in Figure 9. More specifically, friction pads with the structures of extended honeycombs (Figure 9d) and car-tire-like (Figure 9e) exhibited lower stress concentrations, uniformly distributing the applied loads. In addition, the commercial friction pad (Figure 9a) and the friction pad with honeycomb structure (Figure 9b) revealed similar mechanical performance in terms of equivalent von Mises stress concertation showing a maximum von Mises stress value of around 170 MPa. On the other hand, the friction pad with speckled honeycomb structure (Figure 9c) concentrated the highest stress reaching 212 MPa, due to the localized high stress concentration regions derived from its honeycomb spots. It is worth noting that the main stress concentration regions are observed around the support holes of the part, indicating the increased applied loads on the bolts/screws. Finally, it is clear from the numerically extracted mechanical behavior of the designed friction pads that the metal 3D-printed structures are capable of withstanding the friction loads providing promising results for further experimental examination. In conclusion, the higher coefficient of friction configurations exhibited optimal stress distribution, thereby preventing abrupt failure and ensuring more uniform load distribution across braking pads, as revealed by the conducted FEAs.

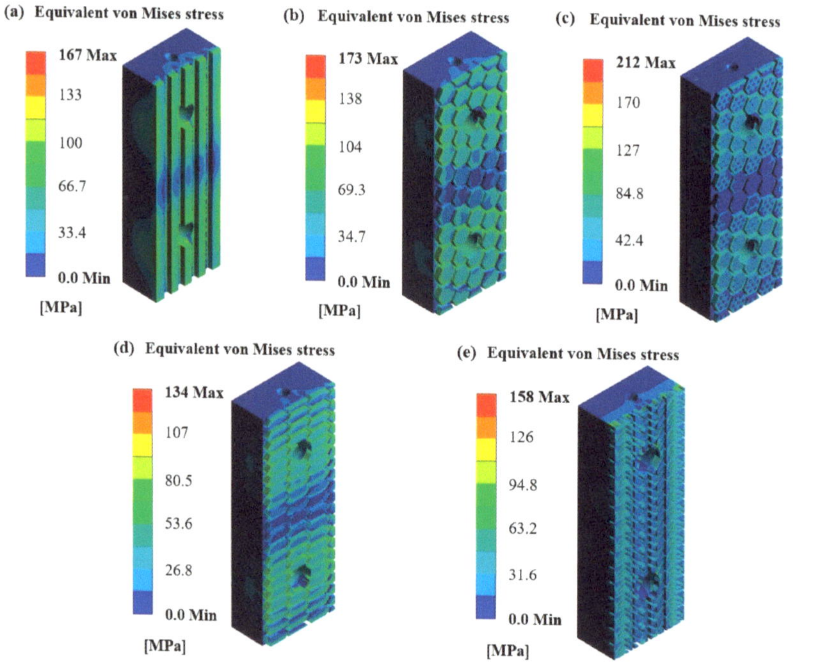

Figure 9. Equivalent von Mises contours for pad structures of: (**a**) Standard (**b**) Honeycomb; (**c**) Speckled honeycomb; (**d**) Extended honeycomb, and (**e**) Car-tire-like.

4. Conclusions

In the current study, elevator progressive safety gear pads with bio-inspired contact surface structures were designed and developed in order to increase their durability and efficiency. More specifically, four different friction pads with embedded bio-inspired surface lattice structures were designed on the template of the conventional pad, namely the honeycomb, speckled honeycomb, extended honeycomb, and car-tire-like structures. Then, these designs were additively manufactured utilizing the SLM 3D printing technique in order to assess their compatibility with the existing system and experimentally evaluate their tribological behavior. The results revealed that the pads with high contact surface area, i.e., car-tire-like and extended honeycomb structure, possess high CoF with 0.549 and 0.459, respectively. Moreover, in the context of this paper, FEMs were developed according to the tribological behavior of each designed friction pad. The results of the numerical analyses showed that the friction pads with high CoF exhibit lower stress, increasing the durability and efficiency of the part. In addition, all designs were compared with the conventional friction pad and revealed superior performance. Then, through the conducted FEA, the developed friction pads experienced significantly lower stress than the construction material strength paving the way for further experimental evaluation through actual application to elevator progressive safety gear. To conclude, it is expected that the reduced loads and stress that the new designs exhibited will increase the number of uses of the braking pads, increasing their life cycle and reducing the maintenance cost and downtime of an elevator system. In addition, the uniform distribution of the stresses deriving from the applied braking force indicates smoother contact between the pads and the guide rails mitigating the risk of deforming them.

5. Patents

The research study was filled on 22 November 2023 in a patent application to the Hellenic Industrial Property Organization (HIPA) with the application number) 20230100969 and the title: «Hierarchical Biomimetic Surface Pattern on Elevator Friction Pads for Enhanced Dynamic Friction Performance».

Author Contributions: Conceptualization, N.K., E.M.P. and I.K.S.; methodology, N.K. and E.M.P.; software, N.K.; validation, N.K., E.M.P. and K.T.; formal analysis, N.K. and K.T.; investigation, N.K. and I.K.S.; resources, A.K. and C.D.; data curation, N.K. and E.M.P.; writing—original draft preparation, N.K.; writing—review and editing, A.K., C.D. and E.M.P.; visualization, I.K.S.; supervision, D.T. and E.M.P.; project administration, E.M.P. and I.K.S.; funding acquisition, D.T. and E.M.P. All authors have read and agreed to the published version of the manuscript.

Funding: This research received no external funding.

Institutional Review Board Statement: Not applicable.

Informed Consent Statement: Not applicable.

Data Availability Statement: The raw data supporting the conclusions of this article will be made available by the authors on request.

Acknowledgments: This research was partially supported by the project "En3DSyst" (Project code: KMP6-0232019) under the framework of the Action "Investment Plans of Innovation" of the Operational Program "Central Macedonia 2014–2020", which is co-funded by the European Regional Development Fund (ERDF) of the European Union and Greece.

Conflicts of Interest: Ioanna K. Sfampa was employed by the KLEEMANN Group. The remaining authors declare that the research was conducted in the absence of any commercial or financial relationships that could be construed as a potential conflict of interest.

References

1. Macalós, J.P.S.; Rossi, P. Up the stairs, down in the elevator? The asymmetric response of emerging market currencies to the global liquidity cycle. *Camb. J. Econ.* **2022**, *46*, 849–868. [CrossRef]

2. Alexandros, V.; Paschalis, C.; Nikolaos, D.; Nikolaos, K.; Michael, C.; Giorgos, G.; Dimitrios, T.; Stelios, K. Automatic elevator shaft inspection using a multi-sensor measuring system and computer vision techniques. *J. Build. Eng.* **2023**, *82*, 108358. [CrossRef]
3. Ma, X.; Pan, G.; Zhang, P.; Xu, Q.; Shi, X.; Xiao, Z.; Han, Y. Experimental evaluation of braking pad materials used for high-speed elevator. *Wear* **2021**, *477*, 203872. [CrossRef]
4. Pulecchi, T.; Manes, A.; Lisignoli, M.; Giglio, M. Digital filtering of acceleration data acquired during the intervention of a lift safety gears. *Measurement* **2010**, *43*, 455–468. [CrossRef]
5. Janovsky, L. *Elevator Mechanical Design*, 2nd ed.; Ellis Horwood Limited: New York, NY, USA, 1993.
6. Strakosch, G.R. *Courtesy of Otis Elevator, The Vertical Transportation Handbook*, 3rd ed.; John Wiley & Sons, Inc.: Hoboken, NJ, USA, 1998.
7. Multi-Part Document BS EN 81—Safety Rules for the Construction and Installation of Lifts. Lifts for the Transport of Persons and Goods. Available online: https://knowledge.bsigroup.com/search?productType=standard&query=BS%20EN%2081&type=products (accessed on 3 June 2024).
8. Peng, Q.; Li, Z.; Yuan, H.; Huang, G.; Li, S.; Sun, X. A Model-Based Unloaded Test Method for Analysis of Braking Capacity of Elevator Brake. *Adv. Mater. Sci. Eng.* **2018**, *2018*, 8047490. [CrossRef]
9. Banik, A.; Tan, K. Dynamic Friction Performance of Hierarchical Biomimetic Surface Pattern Inspired by Frog Toe-Pad. *Adv. Mater. Interfaces* **2020**, *7*, 2000987. [CrossRef]
10. Ivanović, L.; Vencl, A.; Stojanović, B.; Marković, B. Biomimetics Design for Tribological Applications. *Tribol. Ind.* **2018**, *40*, 448–456. [CrossRef]
11. Duralast Products Corporation. Automotive Brake Parts. Available online: https://www.duralastparts.com/brakes.html (accessed on 3 June 2024).
12. Giarmas, E.; Tsakalos, V.; Tzimtzimis, E.; Kladovasilakis, N.; Kostavelis, I.; Tzovaras, D.; Tzetzis, D. Selective Laser Melting Additive Manufactured H13 Tool Steel for Aluminum Extrusion Die Component Construction. *J. Manuf. Process.* 2024, *accepted*.
13. *ASTM E8/E8M-22*; Standard Test Methods for Tension Testing of Metallic Materials. ASTM International: West Conshohocken, PA, USA, 2024.
14. Kladovasilakis, N.; Charalampous, P.; Tsongas, K.; Kostavelis, I.; Tzovaras, D.; Tzetzis, D. Influence of Selective Laser Melting Additive Manufacturing Parameters in Inconel 718 Superalloy. *Materials* **2022**, *15*, 1362. [CrossRef] [PubMed]
15. Kladovasilakis, N.; Charalampous, P.; Kostavelis, I.; Tzetzis, D.; Tzovaras, D. Impact of metal additive manufacturing parameters on the powder bed fusion and direct energy deposition processes: A comprehensive review. *Prog. Addit. Manuf.* **2021**, *6*, 349–365. [CrossRef]
16. *ASTM ISO/ASTM52921-13*; Standard Terminology for Additive Manufacturing-Coordinate Systems and Test Methodologies. ASTM International: West Conshohocken, PA, USA, 2013.
17. *ISO 21920-2:2021*; Geometrical Product Specifications (GPS)—Surface Texture: Profile. ISO: Geneva, Switzerland, 2021.
18. *ISO 25178-2:2021*; Geometrical Product Specifications (GPS)—Surface Texture: Areal. ISO: Geneva, Switzerland, 2021.
19. Borawski, A. Common Methods in Analysing the Tribological Properties of Brake Pads and Discs—A Review. *Acta Mech. Autom.* **2019**, *13*, 189–199. [CrossRef]
20. Lyu, Y.; Bergseth, E.; Wahlström, J.; Olofsson, U. A pin-on-disc study on the tribology of cast iron, sinter and composite railway brake blocks at low temperatures. *Wear* **2019**, *424–425*, 48–52. [CrossRef]

materials

MDPI

Article

Effect of "ColdArc" WAAM Regime and Arc Torch Weaving on Microstructure and Properties of As-Built and Subtransus Quenched Ti-6Al-4V

Anna Zykova, Nikolai Savchenko, Aleksandra Nikolaeva, Aleksander Panfilov, Andrey Vorontsov, Vyacheslav Semenchuk, Denis Gurianov, Evgeny Kolubaev and Sergei Tarasov *

Institute of Strength Physics and Materials Science of the Siberian Branch of the RAS, pr. Academicheskiy, 2/4, Tomsk 634055, Russia; zykovaap@mail.ru (A.Z.); savnick@ispms.ru (N.S.); nikolaeva@ispms.ru (A.N.); alexpl@ispms.ru (A.P.); vav@ispms.ru (A.V.); svm_70@ispms.ru (V.S.); gurianov@ispms.ru (D.G.); eak@ispms.ru (E.K.)

* Correspondence: tsy@ispms.ru; Tel.: +7-3822-286-815

Abstract: Defect-free thin-walled samples were built using wire arc additive manufacturing (WAAM) combined with the "coldArc" deposition technique by feeding a Ti-6Al-4V welding wire and using two deposition strategies, namely with and without the welding torch weaving. The microstructures formed in these samples were examined in relation to mechanical characteristics. The arc torch weaving at 1 Hz allowed us to interfere with the epitaxial growth of the β-Ti columnar grains and, thus, obtain them a lower aspect ratio. Upon cooling, the $\alpha/\alpha'+\beta$ structure was formed inside the former β-Ti grains, and this structure proved to be more uniform as compared to that of the samples built without the weaving. The subtransus quenching of the samples in water did not have any effect on the structure and properties of samples built with the arc torch weaving, whereas a more uniform grain structure was formed in the sample built without weaving. Quenching resulted also in a reduction in the relative elongation by 30% in both cases.

Keywords: wire arc additive manufacturing; "coldArc"; deposition strategy; quenching; microstructure; mechanical properties

Citation: Zykova, A.; Savchenko, N.; Nikolaeva, A.; Panfilov, A.; Vorontsov, A.; Semenchuk, V.; Gurianov, D.; Kolubaev, E.; Tarasov, S. Effect of "ColdArc" WAAM Regime and Arc Torch Weaving on Microstructure and Properties of As-Built and Subtransus Quenched Ti-6Al-4V. *Materials* **2024**, *17*, 2325. https://doi.org/10.3390/ma17102325

Academic Editors: Nikolaos Kladovasilakis, Konstantinos Tsongas and Dimitrios Tzetzis

Received: 17 April 2024
Revised: 10 May 2024
Accepted: 11 May 2024
Published: 14 May 2024

1. Introduction

Wire-arc additive manufacturing (WAAM) is a layer-by-layer metal deposition process using wire as a source material and an arc discharge as an energy source [1,2]. This technology is considered economical and efficient, as well as suitable for the fabrication of large-sized metal products [1] due to its manufacturability, relatively high speed of metal deposition, and low cost of equipment compared to other additive manufacturing methods [2]. Titanium alloys are widely used in various industries, such as aerospace, energy, nuclear, and biomedicine due to their excellent combination of strength and ductility [1,3–5]. At the same time, the use of traditional subtractive metalworking methods to produce shaped titanium alloy parts means a significant increase in the production costs. The development of additive technologies offers an alternative way for fabricating near net-shape machine parts from using a layer-by-layer method, growing them from a melt pool. At the same time, the epitaxial growth of primary columnar β-Ti grains is observed with almost all additive manufacturing methods, subsequently leading to the formation of large α/β colonies with wide α-laths that contribute to the anisotropy of properties and reduce the strength.

During the WAAM process, solidification always occurs through the growth of primary grains from the melt pool bottom under the conditions of a temperature gradient and heat removal through the substrate [6–9]. The relevant studies have shown that the primary β-Ti grains of Ti-6Al-4V alloy produced by WAAM are similar to those obtained using laser [10–12] and electron beam additive manufacturing methods [13–16]. Thus, almost all additive methods that utilize fusion of the source materials give a primary cast structure characterized by large columnar β-Ti grains (several millimeters wide and almost the entire height of the product), in which, at low cooling rates, decomposition occurs with the formation of grain-boundary α -Ti grains and α/β colonies, characterized by low strength and high ductility. With a further decrease in the cooling rate or under the influence of cyclic reheating, the β-laths can experience further decomposition with the formation of α/α'' and α/α_2 structures [17].

Higher cooling rates are achieved in WAAM as compared to those in wire electron beam additive manufacturing that may cause the transformation of α/β structures into α/α' ones, and such a transformation may increase strength but simultaneously impair the ductility [18,19]. When additively forming a Ti-6Al-4V part, prolonged heating can also lead to high residual stresses [20]. Thus, products obtained via additive methods are characterized by either an insufficient level of strength associated with the formation of large primary β-Ti grains and their decomposition into coarse α/β colonies, or low plasticity due to the formation of α/α' colonies. It should be added that the in situ refinement of primary β-Ti grains is possible by breaking the dendrites growing in the melted pool by intensifying the liquid metal stirring by means of mechanical vibrations, sonication (cavitation), and mechanical interpass impact treatment.

Another specificity of the additively manufactured materials is that successive reheating of the as-deposited layers occurs in them, leading to the formation of heat-affected bands [17], in which the as-deposited microstructures are subjected to extra structural and phase transformations such as the $\beta \rightarrow \alpha/\alpha''$ and $\beta \rightarrow \alpha/\alpha_2$. Mending such a structural inhomogeneity may require post-deposition heat treatment.

The modification of deposition strategies may be one of the approaches to interfere with the epitaxial growth of primary columnar β-Ti grains. Plasma or welding torch weaving is an effective way of optimizing the geometry of welding beads and increasing the deposition rate in WAAM [21,22]. In particular, using larger weaving amplitudes in combination with a low torch speed and low wavelength allows for the production of walls of larger dimensions [23]. A novel approach to weaving was developed by Ni et al. [24] for WAAM on curved surfaces that allowed for the improvement of their dimensional accuracy. It was also shown [25] that plasma torch weaving had a positive effect on the structure and strength of reheated zones in the multi-pass plasma-transferred arc Fe-Cr-Mo-C coatings.

Among numerous publications devoted to WAAM on Ti-6Al-4V, there are ones dedicated to studying the effect of different deposition strategies. For instance, Zhou et al. studied the effect of co-directional and reciprocating deposition strategies on structure and mechanical strength, but no weaving was applied in the deposition process [26]. A circular weaving deposition strategy was applied for wire additive manufacturing of Ti-6Al-4V with the use of the cold metal transfer technique [27]. Unfortunately, no data were reported that could shed light on the effect of low-frequency weaving. No literature sources were, however, discovered that reported on a direct study of the effect of arc or plasma torch weaving in WAAM on the structure and characteristics of as-built Ti-6Al-4V.

It may happen, however, that the use of arc torch oscillations in WAAM has positive effect on the structural homogeneity of the WAAM on Ti-6Al-4V.

Alternative solutions may be microalloying the melt pool to extend the constitutional undercooled zone and, thus, increase the probability of homogeneous grain nucleation [28] or intensifying the heterogeneous grain nucleation by adding refractory inoculation particles [29]. The first two approaches involve increased design complexity, while admixing alloying and inoculation additives is not always desirable from the viewpoint of maintaining the chemical composition.

Numerous studies have been carried out to identify the mechanisms determining the microstructural evolution during WAAM and post-heat treatment. For example, the microstructure and mechanical properties of Ti-6Al-4V obtained by the WAAM method were studied after applying five types of the subtransus heat treatment (SHT), from temperatures up to 950 °C [30]. It was found that all five SHT types served to improve both the ultimate strength and ductility. The effect of SHT and hot isostatic pressing on the strength and ductility of Ti-6Al-4V after WAAM was investigated [31]. It was found that various thermal processes directly affect the mechanical properties through the dependence of the strength on the α-grain size [31]. However, it was reported [32] that no significant differences in microstructural or mechanical properties were found after the heat treatment of Ti6Al4V in vacuum (at 720 °C for 120 min), in air, or in argon. All the heat-treated specimens demonstrated both yield and tensile strengths that met the requirements of the AMS4928 standard. A number of sources also indicated the refinement of β-grains, the modification of texture, and residual stresses relieving in the WAAM products after rolling under high pressure [6,8] or sonication [33,34].

As mentioned above, the microstructure of the alloy, including the grain size and morphology, is significantly dependent on the thermal history during the WAAM. In other words, if the thermal conditions during deposition are properly controlled, both the desired microstructures and acceptable mechanical properties can be achieved. Varying the cooling rate by means of interpass cooling with compressed CO_2 showed a slight improvement in tensile strength and a slight decrease in elongation due to the formation of the α'-phase [35,36]. The Ti-6Al-4V wall was obtained by the WAAM with the use of cold metal transfer (CMT) technique and was then subjected to heat treatment [37]. It was shown [37] that the as-built Ti-6Al-4V wall had a microstructure consisting of acicular α'-martensite with a small amount of lamellar α/β colonies and showed acceptable hardness and tensile strength. The martensitic α'-Ti phase can be eliminated by annealing at 900 °C for 4 h and cooling with the furnace. At the same time, the tensile strength of the heat-treated sample was significantly lower than the tensile strength of the original sample [37].

Thus, thermal post-processing of an additively manufactured sample is a realistic route to provide the desired combination of strength and ductility, but it will require additional costs and time. However, it may be more expedient to control the product manufacturing process using techniques that ensure strictly controlled heat input, such as, for example, "coldArc", which uses a power source with real time control of the arc discharge ignition and metal transfer [38,39]. The interplay between WAAM process parameters and deposition strategies which utilize the arc (plasma) torch oscillations is very complicated and multifactorial [23,40]. The use of special WAAM deposition techniques such as CMN or "coldArc" makes this relationship even more complicated, but it still needs further investigations.

In relation to the above-discussed and other papers about using the WAAM process on Ti6Al4V, this work combined WAAM deposition techniques such as "coldArc" with arc torch weaving and evaluated their effect on the grain structure. Thermal post-processing of the samples was carried out to study the necessity of such a procedure for samples obtained by the above-described methods, as well as for understanding the evolution of the microstructure and mechanical properties.

The objective of this study was to elucidate the effect of arc torch weaving on the microstructure and characteristics of the WAAM Ti-6Al-4V samples obtained with the use of "coldArc" technology.

2. Materials and Methods

Welding wire of Ti-6Al-4V, with a diameter of 1.6 mm, was used with the following chemical composition, wt.%: 4.1 ± 0.5 Al, 3.19 ± 0.2 V, Ti—balance. The wire was melted by the arc discharge in a flow of shielding gas (He 99.8%) and then transferred into a melted pool on a 2.5 mm thick rectangular substrate made of commercial pure titanium. The wire arch additive manufacturing (WAAM) was carried out using a 6-axis industrial robot FANUC Robot ARC Mate 100iD (FANUC Europe Corporation, Luxemburg] with a welding torch and an EWM Titan XQ R 400 Pulse inverter power supply (EWM GmbH, Mundersbach, Germany) with "coldArc" technology. The WAAM parameters are presented in Table 1.

Table 1. The "coldArc" WAAM parameters.

Parameter	Magnitude
Mean arc current, A	120
Mean arc voltage, V	14.5
Torch speed, m/min	0.25
Wire feed rate, m/min	3.2
Shielding gas flow rate, l/min	15
Interpass idle time, s	10

The "coldArc" technology allows for the welding process to be carried out with the reduced heat input due to the in situ control of the arc ignition process [41]. It was not possible to calculate the actual heat input due to the lack of parameters and characteristic times of short-circuit and arcing modes. Thus, the nominal value of heat input was calculated the way it is for a conventional arc welding. It may be unnecessary to note here that such an assumption cannot have any effect on the experimental results. Using the WAAM method, two model walls were built from the Ti-6A-4V wire, using different layer deposition strategies: Ti-6Al-4V-0 and Ti-6Al-4V-1 samples (Table 2) were obtained without and with the arc torch weaving, respectively. The weaving amplitude and frequency were 2 mm and 1 Hz (Figure 1). To minimize the distortion of the wall's geometry, the welding torch was turned in the opposite direction after the deposition of each layer. The angle of inclination of the torch relative to the substrate was 10°. Then, the welding torch was raised to the height of the deposited layer, after which the printing direction was changed to the opposite (Figure 1). The chemical composition of the obtained samples is presented in Table 2.

Table 2. Elemental compositions of the "coldArc" WAAM samples.

Layer Deposition Strategy	Zone	Element, wt.%		
		Ti	Al	V
No weaving (Ti-6Al-4V-0)	Bottom	Bal.	3.50 ± 0.57	3.04 ± 0.15
	Middle	Bal.	3.82 ± 0.60	3.09 ± 0.15
	Top	Bal.	4.01 ± 0.62	3.23 ± 0.15
Weaving (Ti-6Al-4V-1)	Bottom	Bal.	3.67 ± 0.56	2.94 ± 0.15
	Middle	Bal.	4.06 ± 0.56	3.03 ± 0.15
	Top	Bal.	4.27 ± 0.63	3.12 ± 0.15

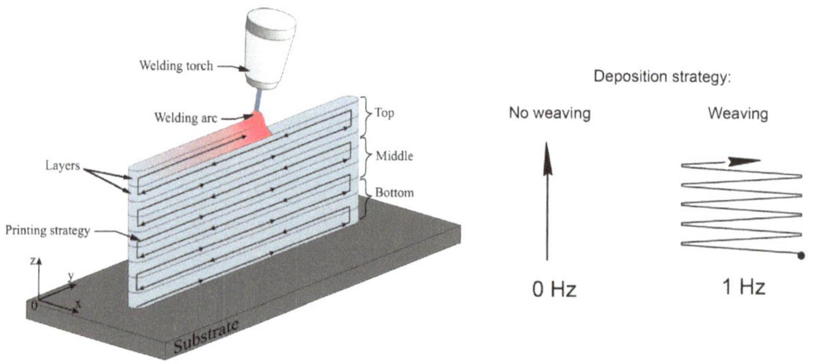

Figure 1. Schematic representation of "coldArc" WAAM on Ti-6Al-4V combined with no weaving/weaving layer deposition strategies.

The as-built walls were sectioned into samples that were subjected to the traditional sample preparation procedure, including the stages of grinding with sandpaper (Al_2O_3) and polishing with diamond pastes (14/10, 3/2, and 1/0 grit). To identify microstructure components, the polished surfaces of the samples were subjected to chemical etching with the reagents 2.5 HNO_3 + 1 HCl + 1 HF + 95 mL H_2O and 4 HF + 6 HCl + 8 HNO_3 + 82 mL H_2O. The macro- and microstructure of the samples were studied using optical microscopy (OM, Altami Met 1S microscope (Altami Ltd., Saint-Petersburg, Russia) and Axiovert 200MAT inverted reflected light microscope(Carl Zeiss AG, Oberkochen, Germany) and scanning electron microscopy (SEM, Thermo Fisher Scientific Apreo S LoVac (Thermo Fisher Scientific, Waltham, MA, USA),) microscope equipped with an energy dispersion spectrometer (EDS)). X-ray diffraction analysis (XRD) was carried out in three zones (bottom, middle, and top) of each wall, using a DRON-7 diffractometer (Bourevestnik, Saint-Petersburg, Russia), $Co_{K\alpha}$ radiation. The level of microstrains (ε) of the WAAM Ti-6Al-4V samples both before and after heat treatment was calculated according to the Williamson–Hall method [42]:

$$\varepsilon = \frac{\beta_{hkl}}{4tg\theta} \tag{1}$$

where β_{hkl} is the full width at half maximum (FWHM), and θ is the Wulf-Bragg angle.

Thin foils were prepared from the thin samples cut from the middle portion of each wall by focused ion beam thinning and then examined by means of a transmission electron microscope (TEM), JEOL-210 (JEOL Ltd., Tokyo, Japan). Uniaxial tensile tests of the dog-bone specimens cut off the corresponding parts of the wall were carried out on a universal testing machine, UTS-110M-100 (Testsystems, Ivanovo, Russia), at room temperature, with a grip speed of 1 mm/min. For uniaxial tensile testing, flat samples were cut with their tensile axes parallel to the OY and with the gauge length dimensions 12 × 2.7 × 1.5 mm.

3. Results

3.1. Macro-, Microstructure, and Phase Composition of the WAAM Ti-6Al-4V Samples

To study structural evolution of the deposited metal as depended on their thermal history, the walls were arbitrarily divided into three parts designated as bottom, middle and top (Figures 2a and 3a). The macrostructures of all the Ti-6Al-4V samples obtained using "coldArc" WAAM and both layer deposition strategies are represented by former columnar-equiaxed primary β-grains, within which the α-phase exists in the form of lathes and Widmanstätten structures (Figures 2b–d and 3b–d).

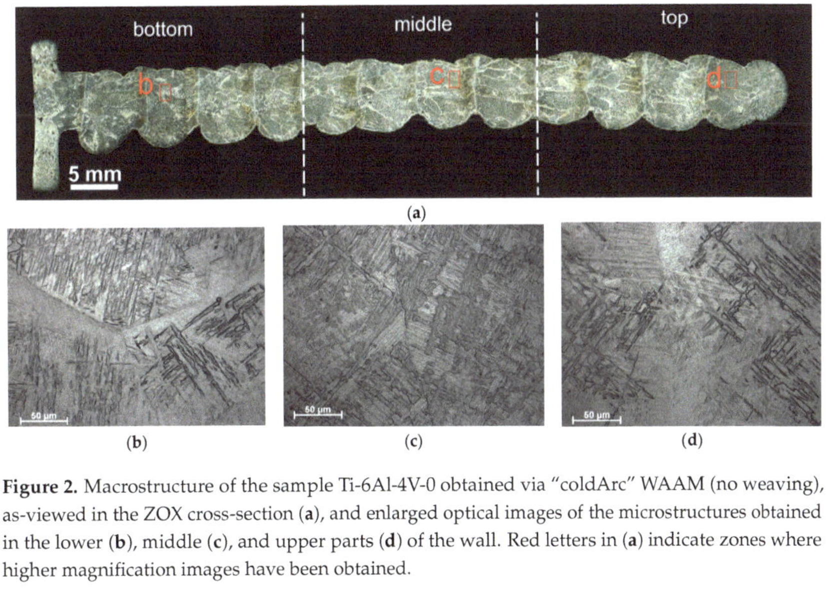

(**a**) (**b**) (**c**) (**d**)

Figure 2. Macrostructure of the sample Ti-6Al-4V-0 obtained via "coldArc" WAAM (no weaving), as-viewed in the ZOX cross-section (**a**), and enlarged optical images of the microstructures obtained in the lower (**b**), middle (**c**), and upper parts (**d**) of the wall. Red letters in (**a**) indicate zones where higher magnification images have been obtained.

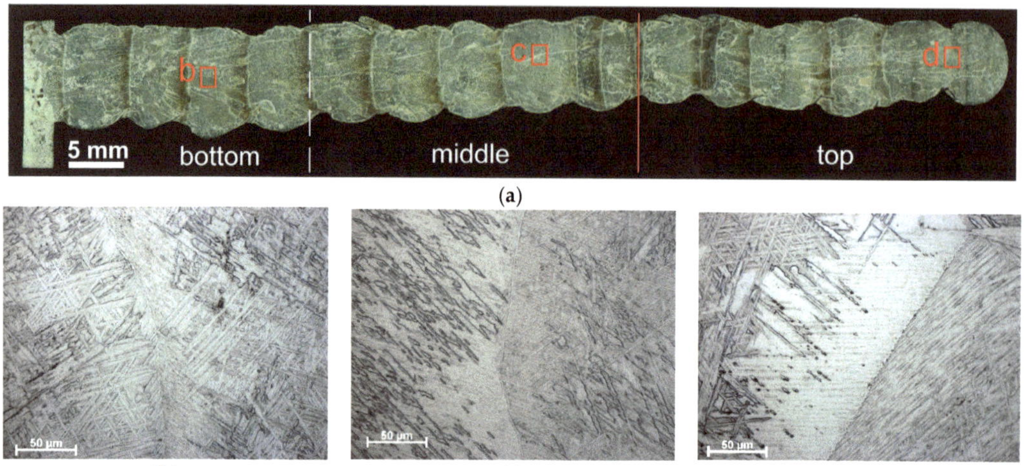

(a) (b) (c) (d)

Figure 3. Macrostructure of the sample Ti-6Al-4V-1 obtained via "coldArc" WAAM (no weaving), as viewed in the ZOX cross-section (**a**), and enlarged optical images of the microstructures obtained in the lower (**b**), middle (**c**), and upper parts (**d**) of the wall. Red letters in (**a**) indicate zones where higher magnification images have been obtained.

It is worth noting that the primary columnar β-Ti grains are characterized by a relatively small mean length (d_{am} = 6 mm) compared to those obtained by WAAM without the "coldArc" technology [1] and other additive methods, where these primary columnar β-grains can reach the length of tens of millimeters and pass through the entire cross-section of the product [11,15,43,44]. The mean size of equiaxed β-grains is almost identical for both layer deposition strategies, i.e., 0.4 ± 0.1 mm and 0.3 ± 0.1 mm in the Ti-6Al-4V-0 and Ti-6Al-4V-1 samples, respectively (Figure 4).

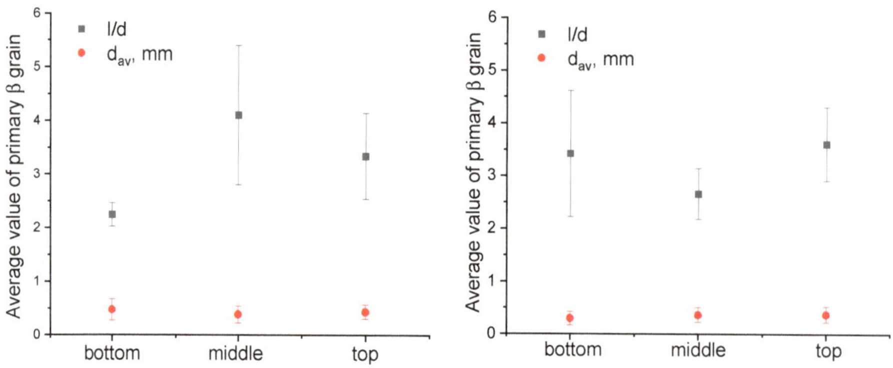

(a) (b)

Figure 4. Mean aspect ratio values for primary columnar and equiaxed β-Ti grains in Ti-6Al-4V-0 (**a**) and Ti-6Al-4V-1 (**b**) samples. l/d—mean aspect ratio of primary columnar β-grains; d_{av}—mean size of equiaxed primary β-grains.

However, the columnar β-grains in Ti-6Al-4V-0 (no weaving) are slightly different in their dimensions, as viewed along the wall's height. The bottom part of the wall contains the columnar grains with a minimum length and width of 0.57 mm and 0.15 mm, respectively, and a maximum of 5.38 mm and 1.37 mm, respectively (Figure 4a). In the middle part of the sample, some columnar grains can reach their maximum sizes of 8 mm in length and 2 mm in width, which leads to an increase in the mean l/d aspect ratio, as well as to the greater scattering of these values (Figure 4a). The top part of the Ti-6Al-4V-0 wall is characterized by isolated high-aspect-ratio grains with a maximum length of up to 7 mm and a width of 1.7 mm.

The use of 1 Hz arc torch transverse oscillations when depositing the Ti-6Al-4V-1 wall resulted in the formation of more uniform grain structures (Figure 3a). The aspect ratio values in the lower, middle, and upper parts of the wall were 3.4 ± 1.2, 2.7 ± 0.7, and 2.6 ± 0.5, respectively; that is, the values were close to each other and fell within the error limits (Figure 4b). In this case, the maximum length and width of the columnar grains were ~9 mm and ~3 mm, respectively.

The macro- and micrographs of the deposited layers in both the Ti-6Al-4V-0 and Ti-6Al-4V-1 samples demonstrate the presence of large primary α-laths, as well as α/β and α/α′ colonies (Figure 5b,c,g,h). Dark equal thickness bands are observed between the deposited layers in all samples that have been formed as a result of the remelting and reheating of the unmelted part of the layer (heat-affected zone) (Figure 5a,f). The microstructure of the reheating zones differs from that of the deposited layer and is characterized by a larger proportion of primary α-laths and α/β colonies located between them (Figure 5d,e,i,j). However, these heat-affected bands look different in the Ti-6Al-4V-0 and Ti-6Al-4V-1 samples. Weaving allowed us to obtain smaller α/β and α/α′ colonies in the heat-affected band (Figure 5i,j) as compared to those in the Ti-6Al-4V-0 sample (Figure 5d,e).

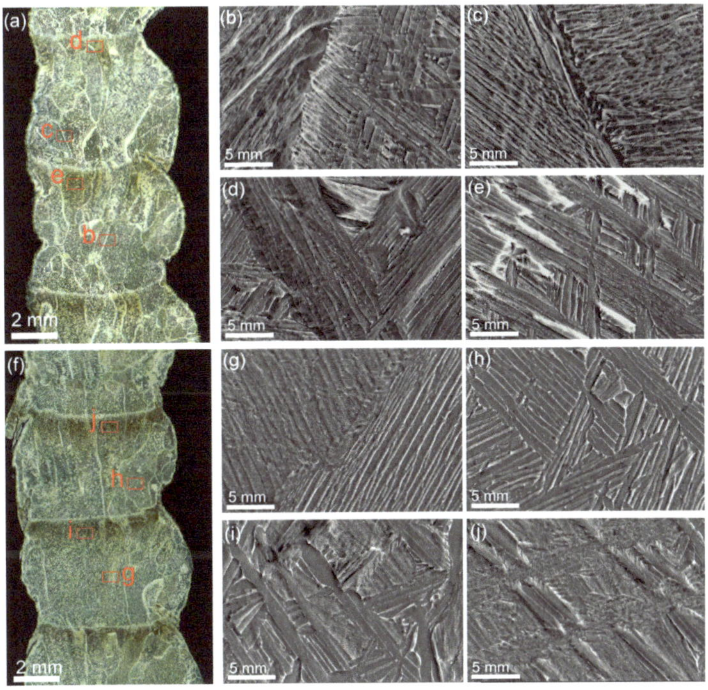

Figure 5. Optical macrographs (**a**,**f**) and SEM images of microstructures in Ti-6Al-4V-0 (**b**–**e**) and Ti-6Al-4V-1 (**g**–**j**). Red letters in (**a**,**f**) indicate areas where higher magnification SEM images have been obtained.

Figure 6 shows the bright-field TEM images of the microstructures in the Ti-6Al-4V-0 and Ti-6Al-4V-1 samples that look identically as well as contain the same structural components such as α-laths, α/β, and α/α′ colonies.

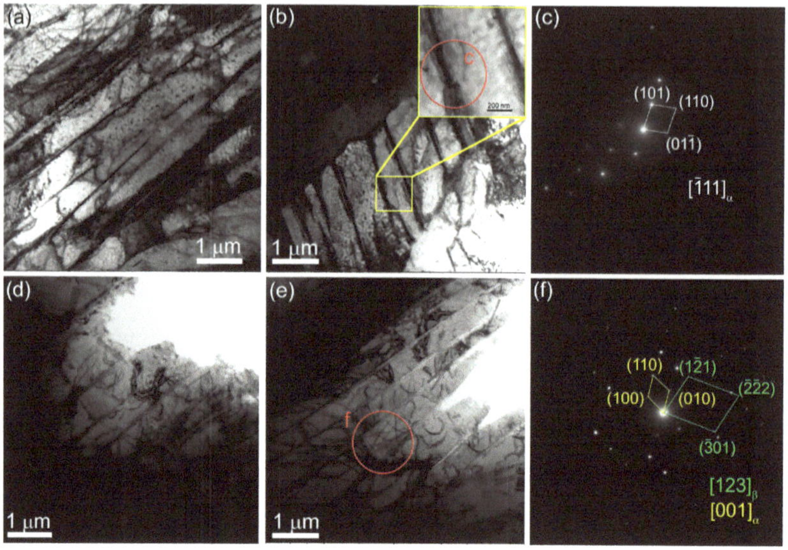

Figure 6. TEM bright-field images (**a**,**b**) and corresponding SAED patterns (**c**) obtained from Ti-6Al-4V-0 (**a–c**) and Ti-6Al-4V-1 (**d–f**) zones. Red letters and circles indicate the areas from which the SAED patterns were obtained and then identified.

The X-ray diffraction patterns of Ti-6Al-4V-0 demonstrate the main reflections from α/α' (HCP) structures, along with weak reflection from $(110)_\beta$ (BCC), and irrespective of the layer deposition strategy. Since α and α' have similar crystal structures and lattice parameters, their X-ray diffraction peaks are indistinguishable, and therefore these structures are designated as α/α' in corresponding X-ray diffraction patterns (Figure 7).

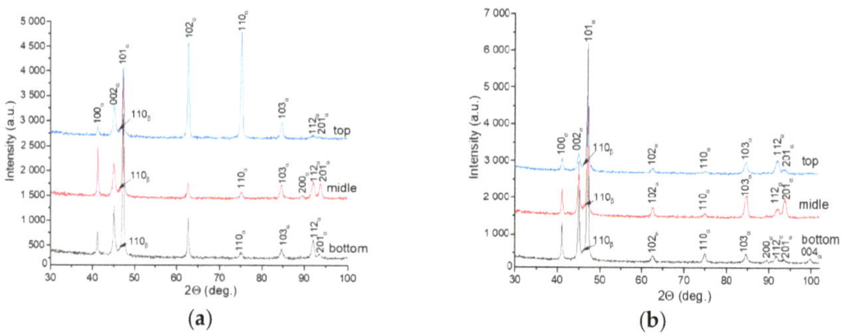

Figure 7. X-ray diffraction patterns obtained from bottom, middle, and top parts of the Ti-6Al-4V-0 (**a**) and Ti-6Al-4V-1 (**b**) walls.

No reflections were observed from other phases, such as, for example, orthorhombic α'' martensite, which appears in the heat-affected bands after wire electron beam additive manufacturing [17]. The XRD patterns of the Ti-6Al-4V-0 sample demonstrate an inverse order of peak intensities that suggests the existence of some texturing (Figure 7a). The $(101)\alpha/\alpha'$ peak has the maximum height in the bottom part of the wall, while both $(102)\alpha/\alpha'$ and $(110)\alpha/\alpha'$ are the predominant ones in the top part of the wall (Figure 7a). For the Ti-6Al-4V-1 sample, peaks (101) of the α/α'-phase stay identical in regard to height throughout the entire height of the wall, which also indirectly confirms the more uniform structure of this sample compared to that of the Ti-6Al-4V-0 sample (Figure 7b).

3.2. The Influence of Heat Treatment on the Structural-Phase State of the "coldArc" WAAM Ti-6Al-4V Samples

The quenching of the Ti-6Al-4V samples from 900 °C followed by cooling in water did not provide any significant microstructural changes (Figure 8). The macrostructure of the WAAM Ti-6Al-4V samples obtained using both layer deposition strategies consisted of columnar-equiaxed primary β-grain boundaries within which the α-phase acquired lamellar, needle-like, and Widmanstätten structures (Figure 8). However, after the subtransus quenching, an increase in the volume fraction of the α' phase was observed. The mean size of equiaxed grains did not change and was identical to that of the as-built samples; for samples Ti-6Al-4V-0 and Ti-6Al-4V-1, the corresponding mean grain sizes are 0.4 ± 0.1 mm and 0.3 ± 0.1 mm, respectively (Figure 9).

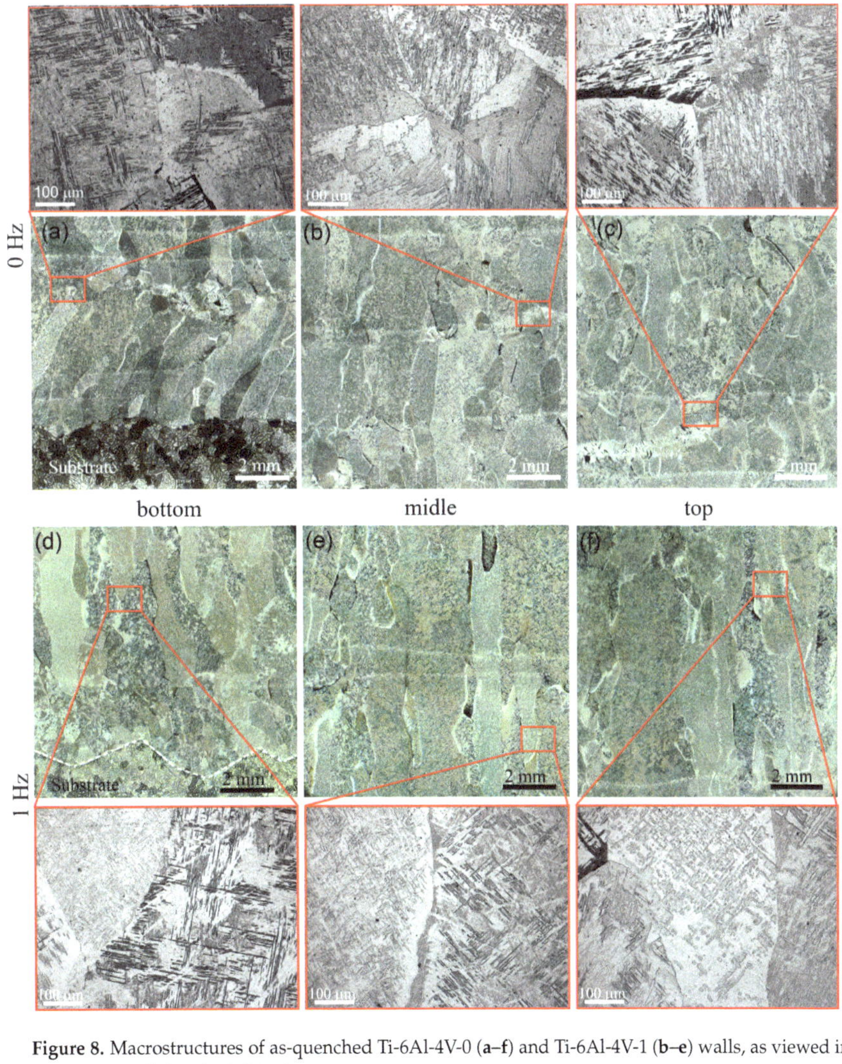

Figure 8. Macrostructures of as-quenched Ti-6Al-4V-0 (**a–f**) and Ti-6Al-4V-1 (**b–e**) walls, as viewed in a section parallel to the ZOX plane.

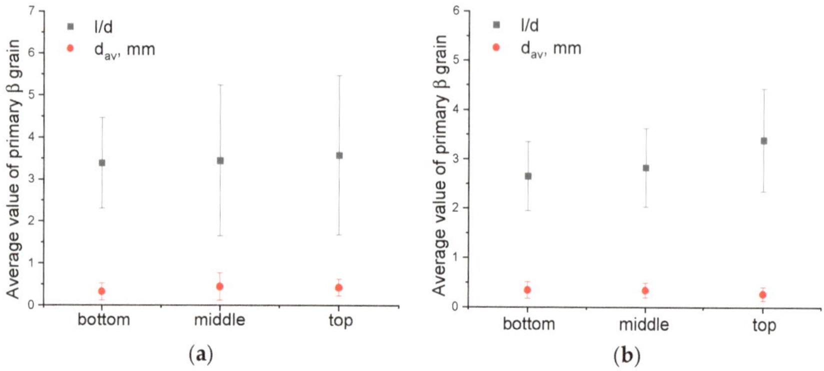

Figure 9. Mean aspect ratio values for primary columnar and size of equiaxial β-Ti grains (d_{av}): Ti-6Al-4V-0 (**a**) and Ti-6Al-4V-1 (**b**) as-quenched samples. l/d—mean aspect ratio of primary columnar β-grains; d_{av}—mean size of equiaxed primary β-grains.

The quenching of the Ti-6Al-4V-0 samples allowed us to obtain more uniform aspect ratio values relating to the primary β-Ti grains (Figure 9a). The mean aspect ratio (l/d) values calculated for the bottom, middle, and top parts of the wall are 3.4 ± 1.1, 3.4 ± 1.1, and 3.6 ± 1.3, respectively (Figure 9a). For the as-quenched Ti-6Al-4V-1 samples, the average aspect ratio values did not change compared to those of the as-built ones, and they were 2.6 ± 0.5, 2.8 ± 0.7, and 3.4 ± 1.2 (Figure 9b).

According to TEM (Figure 10a,b,d,e)), the volume fraction of α′-Ti increased significantly after quenching in both the Ti-6Al-4V-0 and Ti-6Al-4V-1 samples (Figure 10c,f). At the same time, the β-phase was partially retained between the α-plates in the as-quenched Ti-6Al-4V-1 samples (Figure 10e), as confirmed by the corresponding XRD pattern (Figure 11).

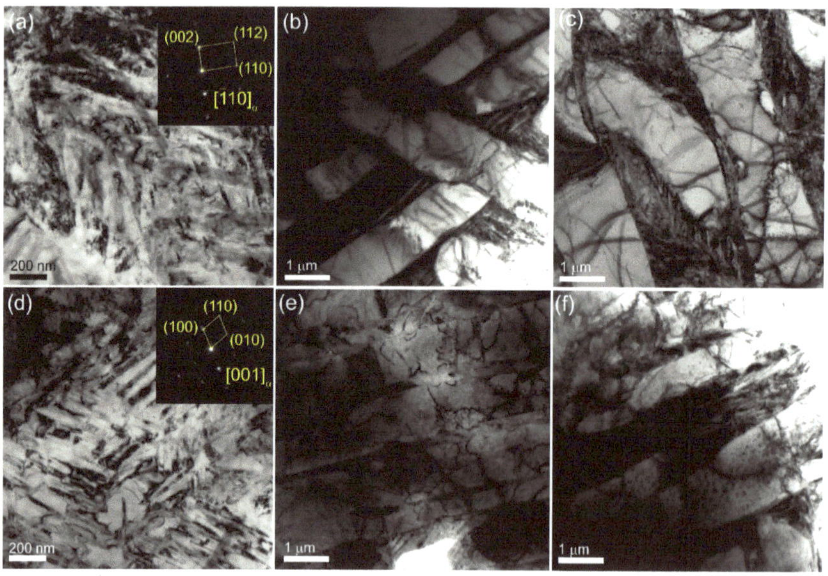

Figure 10. TEM bright-field images of microstructures in as-quenched Ti-6Al-4V-0 (**a–c**) and Ti-6Al-4V-1 (**d–f**) samples.

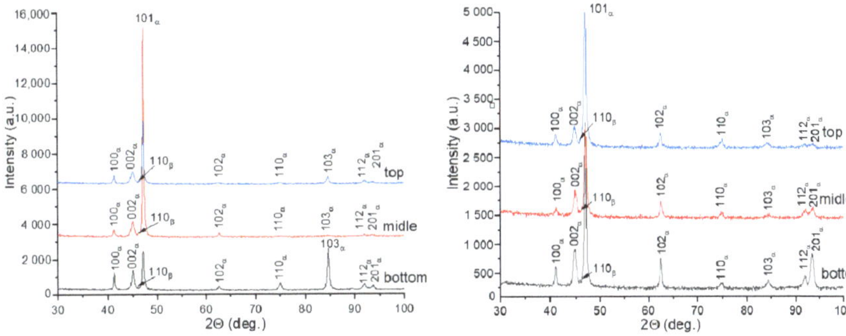

(**a**) (**b**)

Figure 11. X-ray diffractograms obtained from the as-quenched Ti-6Al-4V-0 (**a**) and Ti-6Al-4V-1 (**b**) samples.

The quenching of Ti-6Al-4V-0 samples resulted in the redistribution of the XRD peak intensities compared to those of the as-built samples so that a predominant orientation is observed for the (101)α/α' reflections (Figure 11a). For the as-quenched Ti-6Al-4V-1 samples, the X-ray diffraction patterns were identical to those obtained on the as-built ones with a predominant orientation of (101)α/α' over the entire height of the product.

However, the magnitude of the elastic microstrains of the α/α' lattice depended on the height of the wall (Figure 12). The as-built Ti-6Al-4V-0 samples demonstrated that their elastic microstrain values increased with the wall's height. The quenching of the Ti-6Al-4V-0 samples allowed us to decrease their elastic microstrains, but they retained they dependence on the wall's height (Figure 12a). The as-quenched Ti-6Al-4V-1 samples did not show any linear dependence on the wall's height, thus showing a maximum in the middle part of the wall (Figure 12b).

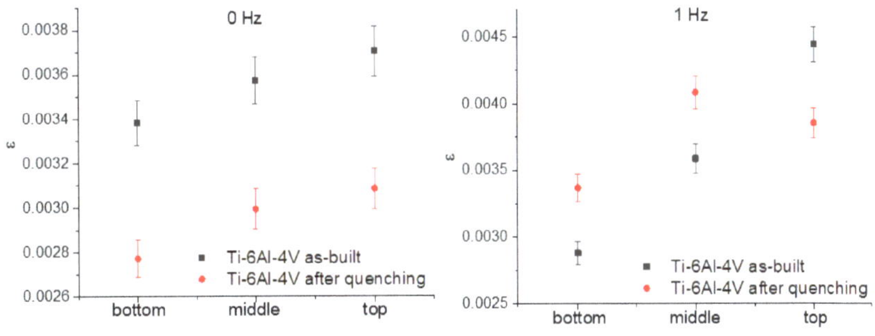

(**a**) (**b**)

Figure 12. Elastic microstrains on the as-built and as-quenched Ti-6Al-4V-0 (**a**) and Ti-6Al-4V-1 (**b**) samples.

3.3. Microhardness

The microhardness profiles for the as-built and as-quenched Ti-6Al-4V-0 and Ti-6Al-4V-1 samples allowed us to observe that both the as-built and as-quenched Ti-6Al-4V-0 samples were characterized by the microhardness number scatter, which was greater than that found in the case of the Ti-6Al-4V-1 sample (Figure 13a,b). The average value of microhardness of the as-built deposited Ti-6Al-4V-0 was 3.4 GPa, which then increased to 3.6 GPa after quenching (Figure 13a). For Ti-6Al-4V-1 samples, the microhardness numbers were slightly lower than those of Ti-6Al-4V-0 and had a smaller scatter (Figure 13b). The average microhardness values of the as-built and as-quenched samples were 3.1 GPa and 3.4 GPa, respectively (Figure 13b).

The increase in microhardness values after quenching for both layer deposition strategies may be due to the increased volume fraction of the α'-Ti phase.

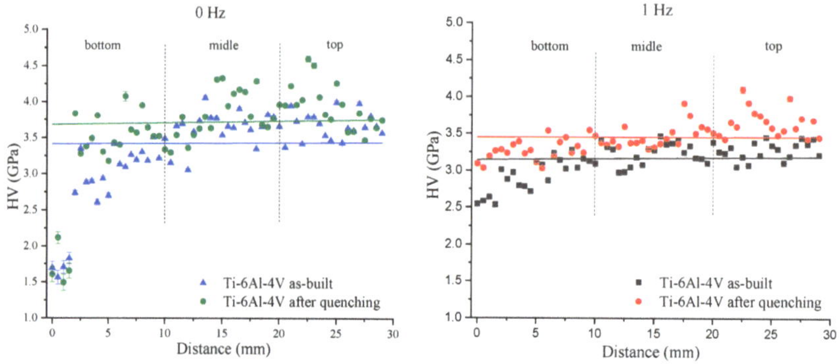

(**a**) (**b**)

Figure 13. Microhardness profiles of as-built and as-quenched Ti-6Al-4V-0 (**a**) and Ti-6Al-4V-1 (**b**) samples.

3.4. Mechanical Strength

Tensile stress–strain curves of as-quenched Ti-6Al-4V-0 and Ti-6Al-4V-1 samples obtained without and with the arc torch weaving show that their mechanical characteristics, such as yield strength (YS), ultimate tensile strength (UTS), and relative elongation (ε), depend on the wall's height (Figure 14). The yield strength values for Ti-6Al-4V-0 in the bottom, middle, and top parts of the product are 531 ± 17, 701 ± 20, and 725 ± 21 MPa, respectively; the same characteristics of the Ti-6Al-4V-1 samples are 560 ± 18, 770 ± 21, and 837 ± 25 MPa (Figure 14a). The bottom parts of the Ti-6Al-4V-0 walls demonstrate their UTS and ε at the levels of 698 ± 19 MPa and 18%, respectively, while for the Ti-6Al-4V-1 samples, they are 643 ± 18 MPa and 14%, respectively (Figure 14a). The UTS values in the middle and upper parts of the as-quenched Ti-6Al-4V-0 and Ti-6Al-4V-1 samples are identical and amount up to 903 ± 28 MPa. In this case, the relative elongation values in the middle and upper parts of the Ti-6Al-4V-0 wall are 11 and 18%, respectively, as compared to the 10 and 8%, respectively, measured on the Ti-6Al-4V-1 samples. Quenching allowed us to increase the tensile strength in the bottom part of both the Ti-6Al-4V-0 and Ti-6Al-4V-1 samples up to 780–800 MPa (Figure 14b). At the same time, the UTS values in the middle and upper parts did not change and were at the level of ~920 MPa, while the relative elongation was reduced to 5% (Figure 14b).

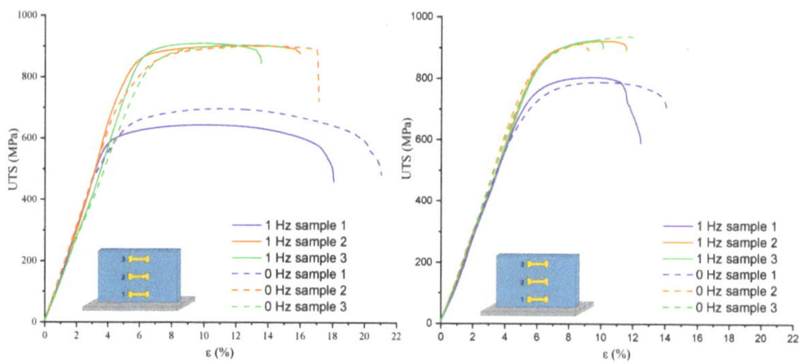

(**a**) (**b**)

Figure 14. Tensile stress–strain curves obtained from as-quenched Ti-6Al-4V-0 (**a**) and Ti-6Al-4V-1 (**b**) samples.

4. Discussion

The Ti-6Al-4V-1 samples were built with the use of arc torch weaving that resulted in smaller β-Ti grains and a more uniform grain structure compared to those of Ti-6Al-4V-0 samples obtained without the weaving. The epitaxial growth of the primary β-Ti grains in the melt pool is determined by the temperature gradient and the associated change in solidification conditions [45,46]. The temperature gradient, nucleation, and growth rate of β-Ti grains are significantly affected by heat input during the additive manufacturing, with a higher temperature gradient typically following less heat input near the liquidus line. Such conditions at the solidification front are accompanied by higher cooling rates, as shown elsewhere [47].

Excessive heat input and possible overheating are known to be problems affecting the stability of the wire arc additive manufacturing process [48,49]. The "coldArc" WAAM features control over the metal transfer via the rapid cycling between arcing and short-circuit modes that requires instantaneous power measurements at the end of each short-circuit mode and immediately before the arc ignition [41]. This ensures a smooth transfer process with a significantly reduced energy input compared to the conventional WAAM. Consequently, the wire arc deposition of the Ti-6Al-4V alloy assisted with "coldArc" technique occurs due to the solidification of a relatively "cold" melt pool, which limits the imposed temperature gradient and, consequently, the columnar growth of the initial β-Ti grains. However, it can be argued that the high heat capacity and low thermal conductivity of the Ti-6Al-4V alloy can lead to significant heat accumulation during the sequential deposition of layers, as well as to the long-term reheating of the previously deposited layers to the temperatures corresponding to the $\alpha \rightarrow \beta$ phase transformation, thus making them capable of supporting the epitaxial growth of β-Ti grains.

Deliberate changes in layer deposition speed and the use of low-frequency transverse arc torch oscillations may interfere with the epitaxial grain growth and promote the formation of a more uniform structure with columnar grains of lower aspect ratio as compared to those solidified without using these oscillations.

According to [49,50], excessively high layer deposition speed can lead to insufficient melting of the wire, increase the nucleation center number in the melt pool and thus promote the formation of smaller equiaxed grains in the neighborhood of the columnar ones. However, this was not observed in the present experiment. On the other hand, combined action of oscillations and use of "coldArc" and CMT processes may contribute to extra grain nucleation centers [51].

The use of forced oscillations with a frequency close to that of the natural frequency of the melt pool (10–200 Hz) may cause a resonance phenomenon in the melt pool with corresponding improvements in stirring the melted pool metal [52–54]. In addition, the continuous oscillations transmitted via a wire feed mechanism may further improve the stirring of the melt pool, promote fragmentation of dendrites, and grain refinement in the dendrite mushy zone [55]. Fragments of dendrites and grains remain in the melt pool and serve as centers of the β-Ti grain nucleation to provide a competitive grain growth mechanism.

The microstructural inhomogeneities discussed above can affect the mechanical characteristics of the WAAM Ti-6Al-4V. In the titanium α/β alloys, the effective dislocation glide length is determined by the average size of α-colonies and the size of α-laths in Widmanstätten structures [56]. Reducing the size of α-laths and α-colonies will increase the amount of high angle boundaries which serve as dislocation barriers and thus increase the dislocation density along the α/β interfaces that will contribute to improvements of the yield strength and hardness. Therefore, as the size of α-lath decreases, an increase in hardness can be expected. However, the α-lath sizes may be determined for both basket-weave and lath-like microstructures, while α-colonies can allow dislocations to easily glide over significant distances and thus reduce the strength properties of the material [57]. These two features can compensate each other and lead to minimal deviations from average hardness values (Figure 9).

This study demonstrates the effect of using the "coldArc" regime in the WAAM on the Ti-6Al-4V on the microstructure and mechanical characteristics that allowed us to obtain the maximum strength and ductility of the alloy compared to those known from the literature. The obtained UTS is higher than that recommended by the ASTM F1108 standard for deformable Ti-6Al-4V (UTS > 860 MPa, Figure 11). This, along with the process's ability to reduce structural and mechanical anisotropy, makes the "coldArc" technology the preferred method for the additive manufacturing on Ti-6Al-4V and other Ti alloys. At the same time, "coldArc" for Ti-6Al-4V requires subsequent heat treatment and cannot be used immediately after manufacturing, because the anisotropy of mechanical properties is not eliminated completely even with the use of welding torch weaving (Figure 11). Considering the current mechanical characteristics of the WAAM Ti-6Al-4V-1, the "coldArc" process combined with the arc torch weaving has great potential for additive manufacturing on this alloy. However, further research is required to optimize the process and develop ideal pre- and post-processing routes for the final product.

5. Conclusions

In the present study, defect-free WAAM walls were grown from Ti-6Al-4V alloy with the use of the "coldArc" technology and welding torch weaving. The microstructural features of the resulting products were investigated to establish the "process–microstructure–property" relationships. Based on the results obtained, the following conclusions were made:

It was shown that "coldArc" WAAM with transverse 1 Hz oscillations of a welding torch resulted in the formation of a more uniform grain structure over the entire height of the wall. This effect may be related to the improved melted pool stirring and agitation, which is the dominant factor in achieving dendrite fragmentation and grain refinement during solidification. The use of weaving in combination with the "coldArc" technique allowed us to reduce the aspect ratio of the primary β-Ti grains in the as-built walls, especially when examining the middle part of the wall. This finding was expected from considering the fact that heat conductivity of the top part of the wall was lower as compared to that of the middle part.

Quenching the samples in water from 900 °C led to the formation of a more uniform grain structure in the Ti-6Al-4V-0 samples built without the welding torch weaving mode. The microstructure (grain size, phase composition, orientation, etc.) and mechanical properties become identical to those found in the as-built Ti-6Al-4V-1 samples obtained with the welding torch weaving.

Quenching had virtually no effect on the structure and properties of samples obtained with the "coldArc" WAAM weaving mode.

After quenching, the relative elongation decreased in both cases. It has been shown that the use of the "coldArc" technology is promising for the additive manufacturing and high-performance production of Ti-6Al-4V products with a high level of mechanical properties if mechanical anisotropy has been eliminated.

Author Contributions: Conceptualization, A.Z. and S.T.; methodology, A.Z., N.S. and S.T.; formal analysis, A.Z, A.N., A.P., N.S., D.G., V.S. and A.V.; investigation, D.G., N.S., A.P., A.V. and V.S.; data curation, V.S., A.Z. and A.N.; writing—original draft preparation, A.Z., S.T. and N.S.; writing—review and editing, S.T.; project administration, E.K.; funding acquisition, E.K. and A.Z. All authors have read and agreed to the published version of the manuscript.

Funding: This research was funded by Government research assignment for ISPMS SB RAS, project FWRW-2024-0001.

Institutional Review Board Statement: Not applicable.

Informed Consent Statement: Not applicable.

Data Availability Statement: The data presented in this study are available on request from the corresponding author.

Acknowledgments: Investigations were carried out using the equipment courtesy of TsKP "Nanotech" (ISPMS SB RAS).

Conflicts of Interest: The authors declare no conflict of interest.

References

1. Lin, Z.; Song, K.; Yu, X. A Review on Wire and Arc Additive Manufacturing of Titanium Alloy. *J. Manuf. Process* **2021**, *70*, 24–45. [CrossRef]
2. Williams, S.W.; Martina, F.; Addison, A.C.; Ding, J.; Pardal, G.; Colegrove, P. Wire + Arc Additive Manufacturing. *Mater. Sci. Technol.* **2016**, *32*, 641–647. [CrossRef]
3. Liu, S.; Shin, Y.C. Additive Manufacturing of Ti6Al4V Alloy: A Review. *Mater. Des.* **2019**, *164*, 107552. [CrossRef]
4. Attar, H.; Ehtemam-Haghighi, S.; Soro, N.; Kent, D.; Dargusch, M.S. Additive manufacturing of low-cost porous titanium-based composites for biomedical applications: Advantages, challenges and opinion for future development. *J. Alloys Compd.* **2020**, *827*, 154263. [CrossRef]
5. Froes, F.H.S.; Gungor, M.N.; Ashraf Imam, M. Cost-affordable titanium: The component fabrication perspective. *JOM* **2007**, *59*, 28–31. [CrossRef]
6. Donoghue, J.; Antonysamy, A.A.; Martina, F.; Colegrove, P.A.; Williams, S.W.; Prangnell, P.B. The effectiveness of combining rolling deformation with Wire–Arc Additive Manufacture on β-grain refinement and texture modification in Ti–6Al–4V. *Mater. Charact.* **2016**, *114*, 103–114. [CrossRef]
7. Syed, A.K.; Zhang, X.; Caballero, A.; Shamir, M.; Williams, S. Influence of deposition strategies on tensile and fatigue properties in a wire + arc additive manufactured Ti-6Al-4V. *Int. J. Fatigue.* **2021**, *149*, 106268. [CrossRef]
8. Colegrove, P.A.; Coules, H.E.; Fairman, J.; Martina, F.; Kashoob, T.; Mamash, H.; Cozzolino, L.D. Microstructure and residual stress improvement in wire and arc additively manufactured parts through high-pressure rolling. *J. Mater. Process. Technol.* **2013**, *213*, 1782–1791. [CrossRef]
9. Wang, F.; Williams, S.; Rush, M. Morphology investigation on direct current pulsed gas tungsten arc welded additive layer manufactured Ti6Al4V alloy. *Int. J. Adv. Manuf. Technol.* **2011**, *57*, 597–603. [CrossRef]
10. Razorenov, S.V.; Garkushin, G.V.; Savinykh, A.S.; Klimova-Korsmik, O.G.; Shalnova, S.A.; Gushchina, M.O. Dynamic Strength of VT6 Titanium Alloy Manufactured by Laser Metal Deposition. *Phys. Mesomech.* **2022**, *25*, 26–32. [CrossRef]
11. Wu, X.; Liang, J.; Mei, J.; Mitchell, C.; Goodwin, P.S.; Voice, W. Microstructures of laser-deposited Ti–6Al–4V. *Mater. Des.* **2004**, *25*, 137–144. [CrossRef]
12. Simonelli, M.; Tse, Y.Y.; Tuck, C. Effect of the build orientation on the mechanical properties and fracture modes of SLM Ti–6Al–4V. *Mater. Sci. Eng. A* **2014**, *616*, 1–11. [CrossRef]
13. Kolubaev, E.A.; Rubtsov, V.E.; Chumaevsky, A.V.; Astafurova, E.G. Micro-, Meso- and Macrostructural Design of Bulk Metallic and Polymetallic Materials by Wire-Feed Electron-Beam Additive Manufacturing. *Phys. Mesomech.* **2022**, *25*, 479–491. [CrossRef]
14. Al-Bermani, S.S.; Blackmore, M.L.; Zhang, W.; Todd, I. The Origin of Microstructural Diversity, Texture, and Mechanical Properties in Electron Beam Melted Ti-6Al-4V. *Metall. Mater. Trans. A* **2010**, *41*, 3422–3434. [CrossRef]
15. Antonysamy, A.A.; Meyer, J.; Prangnell, P.B. Effect of build geometry on the β-grain structure and texture in additive manufacture of Ti6Al4V by selective electron beam melting. *Mater. Charact.* **2013**, *84*, 153–168. [CrossRef]
16. Kalashnikov, K.N.; Chumaevskii, A.V.; Kalashnikova, T.A.; Osipovich, K.S.; Kolubaev, E.A. Defect Formation in Titanium Alloy during Non-stationary Process of Local Metallurgy. *Russ. Phys. J.* **2020**, *63*, 962–967. [CrossRef]
17. Zykova, A.; Panfilov, A.; Vorontsov, A.; Shmakov, V.; Savchenko, N.; Gurianov, D.; Gusarenko, A.; Utyaganova, V.; Krasnoveikin, V.; Tarasov, S. Microstructural and phase evolution of Ti6Al4V in electron beam wire additive manufacturing and on the subtransus quenching and normalization. *Mat. Sci. Eng. A* **2024**, *898*, 146384. [CrossRef]
18. Åkerfeldt, P.; Colliander, M.H.; Pederson, R.; Antti, M.-L. Electron backscatter diffraction characterization of fatigue crack growth in laser metal wire deposited Ti-6Al-4V. *Mater. Charact.* **2018**, *135*, 245–256. [CrossRef]
19. Bruschi, S.; Bertolini, R.; Ghiotti, A. Coupling machining and heat treatment to enhance the wear behaviour of an Additive Manufactured Ti6Al4V titanium alloy. *Tribol. Int.* **2017**, *116*, 58–68. [CrossRef]
20. Li, R.; Xiong, J.; Lei, Y. Investigation on thermal stress evolution induced by wire and arc additive manufacturing for circular thin-walled parts. *J. Manuf. Process.* **2019**, *40*, 59–67. [CrossRef]
21. Ma, G.; Zhao, G.; Li, Z.; Yang, M.; Xiao, W. Optimization strategies for robotic additive and subtractive manufacturing of large and high thin-walled aluminum structures. *Int. J. Adv. Manuf. Technol.* **2019**, *101*, 1275–1292. [CrossRef]
22. He, T.; Yu, S.; Runzhen, Y.; Bo, Z. Oscillating wire arc additive manufacture of rocket motor bimetallic conical shell. *Int. J. Adv. Manuf. Technol.* **2022**, *119*, 6601–6612. [CrossRef]
23. Bultman, J.; Saldana, C. Effects of weave path parameters on the geometry of wire arc additive manufactured features. *Int. J. Adv. Manuf. Technol.* **2023**, *124*, 2563–2577. [CrossRef]
24. Ni, M.; Zhou, Y.; Hu, Z.; Qin, X.; Xiong, X.; Ji, F. Forming optimization for WAAM with weaving deposition on curved surfaces. *Int. J. Mech. Sci.* **2023**, *252*, 108366. [CrossRef]
25. Gnyusov, S.F.; Degterev, A.S.; Tarasov, S.Y. The effect of plasma torch weaving on microstructural evolution in multiplepass plasma-transferred arc Fe-Cr-V-Mo-C coating. *Surf. Coat. Technol.* **2018**, *344*, 75–84. [CrossRef]

26. Zhou, Y.; Qin, G.; Li, L.; Lu, X.; Jing, R.; Xing, X.; Yang, Q. Formability, microstructure and mechanical properties of Ti-6Al-4V deposited by wire and arc additive manufacturing with different deposition paths. *Mater. Sci. Eng. A* **2020**, *772*, 138654. [CrossRef]
27. Farabi, E.; Klein, T.; Schnall, M.; Primig, S. Effects of high deposition rate during cold metal transfer additive manufacturing on microstructure and properties of Ti-6Al-4V. *Addit. Manuf.* **2023**, *71*, 103592. [CrossRef]
28. Zykova, A.P.; Nikolaeva, A.V.; Vorontsov, A.V.; Chumaevskii, A.V.; Nikonov, S.Y.; Moskvichev, E.N.; Gurianov, D.A.; Savchenko, N.L.; Kolubaev E., A.; Tarasov, S.Y. Effect of Copper Content on Grain Structure Evolution in Additively Manufactured Ti-6Al-4V Alloy. *Phys. Mesomech.* **2023**, *26*, 107–125. [CrossRef]
29. Kennedy, J.R.; Davis, A.E.; Caballero, A.E.; Pickering, E.J.; Prangnell, P.B. β grain refinement during solidification of Ti-6Al-4V in Wire-Arc Additive Manufacturing (WAAM). *IOP Conf. Ser. Mater. Sci. Eng.* **2023**, *1274*, 012005. [CrossRef]
30. Wang, J.; Lin, X.; Wang, M.; Li, J.; Wang, C.; Huang, W. Effects of subtransus heat treatments on microstructure features and mechanical properties of wire and arc additive manufactured Ti–6Al–4V alloy. *Mater. Sci. Eng. A* **2020**, *776*, 139020. [CrossRef]
31. Bermingham, M.J.; Nicastro, L.; Kent, D.; Chen, Y.; Dargusch, M.S. Optimising the mechanical properties of Ti-6Al-4V components produced by wire + arc additive manufacturing with post-process heat treatments. *J. Alloys Compd.* **2018**, *753*, 247–255. [CrossRef]
32. Artaza, T.; Suárez, A.; Veiga, F.; Braceras, I.; Tabernero, I.; Larrañaga, O.; Lamikiz, A. Wire arc additive manufacturing Ti6Al4V aeronautical parts using plasma arc welding: Analysis of heat-treatment processes in different atmospheres. *J. Mater. Res. Technol.* **2020**, *9*, 15454–15466. [CrossRef]
33. Gou, J.; Wang, Z.; Hu, S.; Shen, J.; Tian, Y.; Zhao, G.; Chen, Y. Effects of ultrasonic peening treatment in three directions on grain refinement and anisotropy of cold metal transfer additive manufactured Ti-6Al-4V thin wall structure. *J. Manuf. Process.* **2020**, *54*, 148–157. [CrossRef]
34. Yang, Y.; Jin, X.; Liu, C.; Xiao, M.; Lu, J.; Fan, H.; Ma, S. Residual Stress, Mechanical Properties, and Grain Morphology of Ti-6Al-4V Alloy Produced by Ultrasonic Impact Treatment Assisted Wire and Arc Additive Manufacturing. *Metals* **2018**, *8*, 934. [CrossRef]
35. Wu, B.; Pan, Z.; Ding, D.; Cuiuri, D.; Li, H.; Fei, Z. The effects of forced interpass cooling on the material properties of wire arc additively manufactured Ti6Al4V alloy. *J. Mater. Process. Technol.* **2018**, *258*, 97–105. [CrossRef]
36. Ding, D.; Wu, B.; Pan, Z.; Qui, Z.; Li, H. Wire arc additive manufacturing of Ti6AL4V using active interpass cooling. *Mater. Manuf. Process.* **2020**, *35*, 845–851. [CrossRef]
37. Gou, J.; Shen, J.; Hu, S.; Tian, Y.; Liang, Y. Microstructure and mechanical properties of as-built and heat-treated Ti-6Al-4V alloy prepared by cold metal transfer additive manufacturing. *J. Manuf. Process.* **2019**, *42*, 41–50. [CrossRef]
38. Al Zamzami, I.; Davison, B.; Susmel, L. Nominal and local stress quantities to design aluminium-to-steel thin welded joints against fatigue. *Int. J. Fatigue.* **2019**, *123*, 279–295. [CrossRef]
39. Biester, K.; Lammers, M.; Hermsdorf, J.; Kaierle, S. Influence of the laser position in laser-assisted WAAM process on weld bead shape and surface properties. *Procedia CIRP* **2022**, *111*, 190–195. [CrossRef]
40. Zhou, S.; Xie, H.; Ni, J.; Yang, G.; Qin, L.; Guo, X. Metal transfer behavior during CMT-based Wire Arc Additive Manufacturing of Ti-6Al-4V alloy. *J. Manuf. Process.* **2022**, *82*, 159–173. [CrossRef]
41. Goecke, S.F. Low Energy Arc Joining Process for Material Sensitive to Heat: Report—EWM Mündersbach, Germany, 2005; pp. 1–5. Available online: https://studyres.com/doc/5825671/low-energy-arc-joining-process-for-materials-sensitive-to...# (accessed on 10 May 2024).
42. Gorelik, S.S.; Rastorguev, L.N.; Skakov, Y.A. *X-ray and Electron-Optical Analysis*; Metallurgy: Moscow, Russia, 1970; 366p.
43. Brandl, E.; Schoberth, A.; Leyens, C. Morphology, microstructure, and hardness of titanium (Ti-6Al-4V) blocks deposited by wire-feed additive layer manufacturing (ALM). *Mater. Sci. Eng. A.* **2012**, *532*, 295–307. [CrossRef]
44. Wu, X.; Sharman, R.; Mei, J.; Voice, W. Direct laser fabrication and microstructure of a burn-resistant Ti alloy. *Mater. Des.* **2002**, *23*, 239–247. [CrossRef]
45. Gäumann, M.; Bezençon, C.; Canalis, P.; Kurz, W. Single-crystal laser deposition of superalloys: Processing–microstructure maps. *Acta Mater.* **2001**, *49*, 1051–1062. [CrossRef]
46. Rosli, N.A.; Alkahari, M.R.; bin Abdollah, M.F.; Maidin, S.; Ramli, F.R.; Herawan, S.G. Review on effect of heat input for wire arc additive manufacturing process. *J. Mater. Res. Technol.* **2021**, *11*, 2127–2145. [CrossRef]
47. Ou, W.; Knapp, G.L.; Mukherjee, T.; Wei, Y.; DebRoy, T. An improved heat transfer and fluid flow model of wire-arc additive manufacturing. *Int. J. Heat Mass Transf.* **2021**, *167*, 120835. [CrossRef]
48. Ding, D.; Pan, Z.; Cuiuri, D.; Li, H. Wire-feed additive manufacturing of metal components: Technologies, developments and future interests. *Int. J. Adv. Manuf. Technol.* **2015**, *81*, 465–481. [CrossRef]
49. Ma, Y.; Cuiuri, D.; Shen, C.; Li, H.; Pan, Z. Effect of interpass temperature on in-situ alloying and additive manufacturing of titanium aluminides using gas tungsten arc welding. *Addit. Manuf.* **2015**, *8*, 71–77. [CrossRef]
50. Ahn, D.-G. Directed Energy Deposition (DED) Process: State of the Art. *Int. J. Precis. Eng. Manuf. Technol.* **2021**, *8*, 703–742. [CrossRef]
51. Cunningham, C.R.; Flynn, J.M.; Shokrani, A.; Dhokia, V.; Newman, S.T. Invited review article: Strategies and processes for high quality wire arc additive manufacturing. *Addit. Manuf.* **2018**, *22*, 672–686. [CrossRef]
52. Ke, W.C.; Oliveira, J.P.; Cong, B.Q.; Ao, S.S.; Qi, Z.W.; Peng, B.; Zeng, Z. Multi-layer deposition mechanism in ultra high-frequency pulsed wire arc additive manufacturing (WAAM) of NiTi shape memory alloys. *Addit. Manuf.* **2022**, *50*, 102513. [CrossRef]
53. Hirata, Y. Pulsed arc welding. *Weld. Int.* **2003**, *17*, 98–115. [CrossRef]

54. Lin, J.J.; Lv, Y.H.; Liu, Y.X.; Xu, B.S.; Sun, Z.; Li, Z.G.; Wu, Y.X. Microstructural evolution and mechanical properties of Ti-6Al-4V wall deposited by pulsed plasma arc additive manufacturing. *Mater. Des.* **2016**, *102*, 30–40. [CrossRef]
55. Sequeira Almeida, P.M.; Williams, S. Innovative process model of Ti-6Al-4V additive layer manufacturing using cold metal transfer (CMT). In Proceedings of the 21st Annual International Solid Freeform Fabrication Symposium—An Additive Manufacturing Conference, Virtual, 2–4 August 2021; pp. 25–36.
56. Lütjering, G.; Williams, J.C. *Titanium*; Springer: Berlin/Heidelberg, Germany, 2013; 379p, Available online: https://books.google.ru/books?id=wBXsCAAAQBAJ (accessed on 10 May 2024).
57. Tiley, J.; Searles, T.; Lee, E.; Kar, S.; Banerjee, R.; Russ, J.C.; Fraser, H.L. Quantification of microstructural features in α/β titanium alloys. *Mater. Sci. Eng. A* **2004**, *372*, 191–198. [CrossRef]

materials

Article

Improvement of the Technology of Precision Forging of Connecting Rod-Type Forgings in a Multiple System, in the Aspect of the Possibilities of Process Robotization by Means of Numerical Modeling

Marek Hawryluk [1,*], Łukasz Dudkiewicz [1,2], Sławomir Polak [1], Artur Barełkowski [1], Adrian Miżejewski [2] and Tatiana Szymańska [2]

[1] Department of Metal Forming, Welding and Metrology, Wroclaw University of Science and Technology, Lukasiewicza Street 5, 50-370 Wroclaw, Poland; lukasz.dudkiewicz@pwr.edu.pl (Ł.D.); slawomir.polak@pwr.edu.pl (S.P.); artur.barelkowski@pwr.edu.pl (A.B.)
[2] Schraner Polska, Lotnicza Street 21G, 99-100 Łęczyca, Poland; amizejewski@schraner.pl (A.M.); tszymanska@schraner.pl (T.S.)
* Correspondence: marek.hawryluk@pwr.edu.pl

Abstract: The study refers to the application of numerical modeling for the improvement of the currently realized precision forging technology performed on a hammer to produce connecting rod forgings in a triple system through the development of an additional rolling pass to be used before the roughing operation as well as preparation of the charge to be held by the robot's grippers in order to implement future process robotization. The studies included an analysis of the present forging technology together with the dimension–shape requirements for the forgings, which constituted the basis for the construction and development of a thermo-mechanical numerical model as well as the design of the tool construction with the consideration of the additional rolling pass with the use of the calculation package Forge 3.0 NxT. The following stage of research was the realization of multi-variant numerical simulations of the newly developed forging process with the consideration of robotization, as a result of which the following were obtained: proper filling of the tool impressions (including the roller's impression) by the deformed material, the temperature distributions for the forging and the tools as well as plastic deformations (considering the thermally activated phenomena), changes in the grain size as well as the forging force and energy courses. The obtained results were verified under industrial conditions and correlated with respect to the forgings obtained in the technology applied so far. The achieved results of technological tests confirmed that the changes introduced into the tool construction and the preform geometry reduced the diameter, and thus also the volume, of the charge as well as provided a possibility of implementing robotization and automatization of the forging process in the future. The obtained results showed that the introduction of an additional rolling blank resulted in a reduction in forging forces and energy by 30% while reducing the hammer blow by one. Attempts to implement robotization into the process were successful and did not adversely affect the geometry or quality of forgings, increasing production efficiency.

Keywords: hammer precision forging; forging FEM modeling; robotization; technology improvement

Citation: Hawryluk, M.; Dudkiewicz, Ł.; Polak, S.; Barełkowski, A.; Miżejewski, A.; Szymańska, T. Improvement of the Technology of Precision Forging of Connecting Rod-Type Forgings in a Multiple System, in the Aspect of the Possibilities of Process Robotization by Means of Numerical Modeling. *Materials* **2024**, *17*, 1087. https://doi.org/10.3390/ma17051087

Academic Editors: Dimitrios Tzetzis, Konstantinos Tsongas and Nikolaos Kladovasilakis

Received: 30 January 2024
Revised: 15 February 2024
Accepted: 22 February 2024
Published: 27 February 2024

1. Introduction

Connecting rods are elements that join the drive systems in many machines and devices, including petrol chainsaws used, e.g., in the woodworking industry. They constitute an important safety element of the drive system and for this reason, they have to be characterized by high dimension–shape precision, quality, and performance properties [1]. In large lot production, for forgings of the connecting rod type, hot precision forging is applied in open or closed dies, usually on crank and screw presses, as well as hydraulic hammers, often in multiple systems, in order to improve the production efficiency with the preservation of the high product quality [2]. The obtained forgings are subjected to flash trimming as well as some additional procedures (cooling, thermo-chemical treatment, shot peening, finishing treatment through machining, defectoscopic tests, and final inspection). The process of producing connecting rod-type forgings, especially in multiple systems, still constitutes a big challenge and a not entirely solved problem, as, in the die forging processes, a huge role is played by the development of a proper tool construction and working impression geometry as well as selection of the correct technological parameters of the process, which translates to technology efficiency and product quality, with the consideration of tool durability. This is of significant importance in the case when the forgings are required to have an additionally specified structure and hardness, which are obtained as a result of both the forging process and, often, a proper thermal treatment [3]. The most important factors affecting these aspects are the technological parameters, the shape and quality of the tools as well as the number of operations, the geometry of the preform and the slug forging, and also the thermal parameters affecting the tribological conditions [4,5]. Although the forging technology is relatively well-known, the proper preparation of the forgings, especially ones of a complicated shape, will fulfill the precision and quality requirements demanded from the constructors and technologists, as well as operators, to possess extensive knowledge and experience [6]. A certain aid as well as solution in the increase in stabilization and efficiency of production is the introduction of automatization and robotization of the forging processes [7], however, it is an additional challenge, requiring time and financial resources. The literature provides few applications of effective automatization and robotization of die forging processes, which, however, are dedicated to specific types of forgings, and each one requires an individual approach [8,9]. This creates the need to design the technology from scratch, as it requires a redesign of the tools and consideration of the role of the robots' grippers in the process [10]. In the case of hot die forging, one should also consider the abrasive wear of the tools' working impressions during production, which causes material losses, which, in turn, increase the forging volume at the expense of the forging material, which makes the gripping of the forging by the robot significantly more difficult, and this should also be considered in the design of automated forging processes. In the case of forging process automatization, one should especially take into account not only the vibration but also the large elastic deflections of the whole tool sets, which requires the application of compensation in the gripper joints/their docking, or the application of other solutions [11]. Such solutions coupled by means of machine communication enable control of the slide as well as the time of the hold-up for the required time [12,13]. This is also connected with the implementation of additional sensors and detectors, elements of electronics, as well as the application of specialized software, and also the necessity of performing numerous tests and trials [14]. The attempts at introducing automatization and robotization of hot forging processes, in the case of using at least a part of the outdated machinery park, are even more difficult [15,16]. At the same time, at every stage of the technological line, there is a potential risk of the occurrence of a defect/problem lowering the quality of the products or causing unfulfillment of the geometrical or microstructural requirements (improper structure after forging, inappropriate hardness, etc. [17]. Also, many times, the cause of the errors identified at a given stage is an improperly developed technology, also in the aspect of robotization, and/or it not being followed at the earlier stages of the process [18]. As we know, a properly elaborated industrial process of plastic treatment requires the realization of numerous experiments and tests, which is connected with huge costs and a lot of devoted time [19], and still, the most important stage of design and optimization is the final verification of the developed process under industrial condition [20,21]. For this reason, at present, for the analysis, optimization, or elaboration of new technology, engineering programs based on CAD/CAM/CAE are used, which are aided by IT tools [22–24], as well as the very popular numerical modeling, which often constitutes an independent tool [25,26]. The simultaneous application of many methods and techniques aiding the design, simulation, and production enables a global and complex approach to the given problem [27,28]. The available literature provides many studies and articles referring to the use of numerical modeling techniques for the analysis and optimization of plastic-forming processes. In such a case, numerical modeling based on FEM/FVM [29] is mostly applied for the determination of the optimal shape and dimensions of the slug forging, as well as the material flow and filling of the impression, and also the temperature field, both in the forging and the tools, or for the determination of the deformation distribution, and even the grain size after the process [30]. The studies [31,32] apply FEM for a complex analysis of the forging process in order to improve the analyzed technology. In turn, in the works [33,34], FEM was used for an analysis of the process of flash trimming. The currently used calculation packages are equipped with more and more functions enabling an even better and more thorough analysis of the metal forming processes [35–37], making it possible, e.g., to analyze the wear of the dies and forging punches (Forge, QFORM, Simufact, Deform) [38,39]. More and more often, in FEM [40,41], special functions for flaw detection are applied [42]. The application of such functions by the user makes it possible to significantly shorten the implementation time of a new project and limit the errors, e.g., during the design of new instrumentation taking into account the gripping by the robots' grippers [43]. Although numerical modeling and IT tools [44] significantly change the role and scope of the experiment to the virtual dimension, the real experiment remains, on the one hand, the best and most necessary verification, and on the other hand, it is the most expensive and time-consuming stage of design [45]. Nonetheless, numerical modeling has been and still is a very convenient, fast, and most commonly applied tool for the analysis and optimization of production processes, including the robotization and automatization of die forging processes [46]. What is more, a continuous expansion of such IT tools with new functions and capabilities makes it possible to approach the real experiment even closer. The additional support of the numerical modeling results with measurement techniques (3D scanning, etc.), as well as microstructure examinations, enables a complex analysis of the whole technology, as well as its improvement and development [47,48].

The aim of the research is to construct a proper numerical model of a hot forging process of producing a connecting rod forging in a triple system, in order to improve the current technology, especially in the aspect of introducing an additional rolling pass (to increase efficiency), as well as optimizing the technological parameters of the process and the possibilities of robotization of the precision forging process realized so far.

2. Test Subject and Methodology

In the research, an analysis was performed of a 3-type forging elongated forgings-high (with a stem) with protrusions of a connecting rod type, produced in a triple system. Figure 1a presents the application of this forging product. A CAD model of a single forging is shown in Figure 1b. A dimensioned forging is presented in Figure 1c.

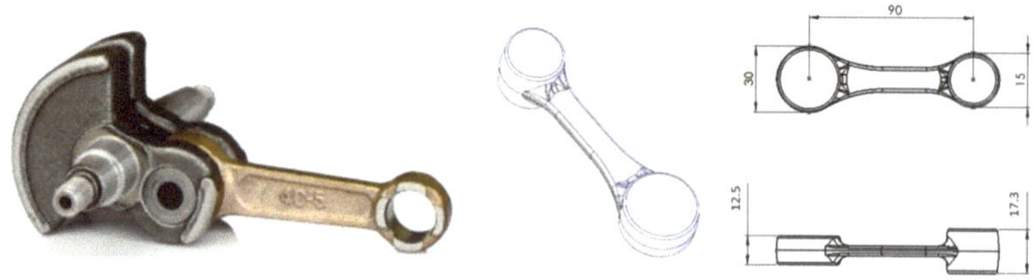

(**a**) (**b**) (**c**)

Figure 1. View of (**a**) a CAD model, (**b**) a photograph of the forgings after the consecutive stages of the manual production process, and (**c**) a single-dimensioned forging.

The current process of producing a connecting rod forging is realized in 2 forging operations: roughing and finishing forging. The charge material is a bar, with a diameter of 16 mm and 280 mm long, made of carburizing steel 13CrMo4-5 (1.7335). The number of details forged in one cycle: 3. The tools (forging dies) are made of 1.2367 steel, and, after the thermal treatment, their hardness is at the level of 53–55 HRC. After being cut, the charge material is heated in an induction furnace, which is followed by forging on a hydraulic hammer with energy of 16 kJ. Whether the connecting rod forgings are properly made is determined by the following requirements: dimension–shape accuracy at the level of 0.1–0.3 mm, mean grain size of 5 according to ASTM, and hardness at the level of 170–200 HB, as well as lack of surface defects or laps; joggle at the level of 0.3 mm is accepted. Technological allowances for this type of connecting rods range from 0.5 mm to 1 mm per side. In the technology implemented so far, the process is realized manually.

In order to achieve the set goal, the following methods and techniques, as well as measurement and testing tools, were used in the realization of the consecutive research and development studies:

- a complex analysis of the forging process with the use of, e.g., a thermovision camera Flir 840 (Wilsonville, OR, USA) and a high-speed camera (Casio Pro Ex-F3, Casio, Tokyo, Japan), as well as a macroscopic analysis of the tools and the forging defects by means of a camera Cannon EOSx 60D (Cannon, Tokyo, Japan).
- three-dimensional scanning Atos Core 135 (GOM, Braunschweig, Germany) structured light scanner, equipped with two 5MPix CCD cameras (GOM, Braunschweig, Germany) (resolution 2448 × 2050 pixels) with a measuring field of 135 × 100 mm, working distance of 170 mm, and physical point distance of 0.05 mm for a single scan.
- based on the current specification sheet, development of CAD models of a ready forging as well a tool (additional rolling pass) with the consideration of the aspects of robotization (increased length of the charge through the use of the so-called "tickworm" enabling a double-sided grip by the manipulators' grippers) by means of the program Catia V6R20 by Dassault, Paris, France.
- based on the above information, a numerical model was developed and simulations of the improved technology of hot precision forging were made with the use of the calculation package of the QForm program (https://www.qform3d.com/ (accessed on 21 February 2024)) in order to determine the key parameters and physical quantities as well as identify the most important problems.
- modeling of the trajectory of the robots' movement (RobotSudion ABB 2020).
- in order to verify the introduced change and improvements resulting mostly from the numerical modeling, a measurement of the forgings geometry was made in reference to a forging before the changes, as well as an analysis of the obtained microstructure and hardness of the produced forgings.
- microstructural observations (for verification purposes) with the use of a light microscope Leica Dm6000N (Leica Microsystems, Tokyo, Japan). To that end, the die insert was incised along the shorter side to prepare samples for the tests. The grinding and polishing, in order to obtain traditional micro-sections, was conducted on a grinder-polisher Struers 350. For the etching, a picric acid solution was used.
- hardness measurements made by means of a hardness tester LECO LC120 (LECO, St. Joseph, MO, USA);

Figure 2 presents a diagram in order to better understand the test process.

It should be emphasized that a thorough implementation of full robotization combined with automatization includes many more technological and technical aspects, which have not been discussed in this article (e.g., automatization of the loading and temperature segregation, determination of the local key devices, selection of robots, design of the manipulation grippers, assembly and replacement of tools, manner of their installation on the hammer as well as heating and lubrication of the tools in the forging process, etc.).

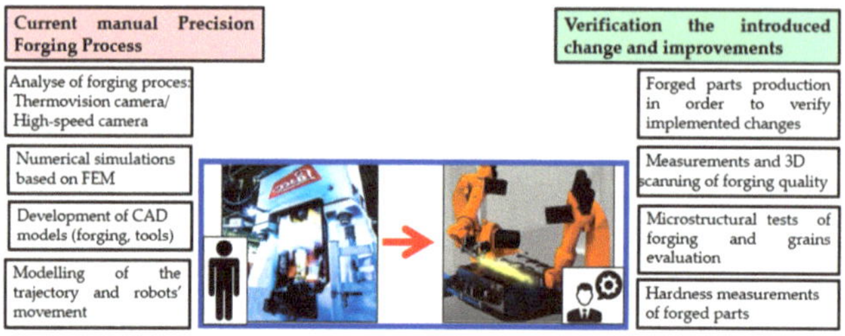

Figure 2. Diagram of the test process.

3. Results and Discussion

The research was divided into a few stages, among which the first three included an analysis of the technology implemented so far in the manual system, as well as the design and modeling of the forging process with the consideration of the aspects of robotization. In turn, the last stage referred to preliminary tests under industrial conditions for verification of the performed research.

3.1. Analysis of the Current Technology in a Manual System

Forgings characterized by an elongated shape with a big difference between the cross-sessions along the length, such as connecting rods, are usually made from a slug forging, in which the material is already preliminarily formed. Forgings can be produced in different ways. The most efficient method is the use of dedicated machines, i.e., forging rolling mills, in which the milling can take place by the method of transverse rolling or periodic rolling. However, in the case of no rolling mill, the forming process can happen directly on the hammer through open or semi-open die forging in the dedicated spot on the die. In the current process of forging a connecting rod, a multiple system is used, in which three forgings are made from a round bar. The forging process takes place in two operations. The first forging operation, i.e., roughing, consists of flattening a cylindrical preform placed on the diameter. This operation involves the highest pressures and material deformations. The second operation is finishing forging, as a result of which we obtain a shape close to that of the ready product. The lubricant is used as a mixture of graphite with water in proportion 1:16. The efficiency of the current technology is 2100 items per shift (700 leaves); the cycle time for the manual process is 19 s. Figure 3 shows the thermograms from the process for the charge, the tools, and the forging.

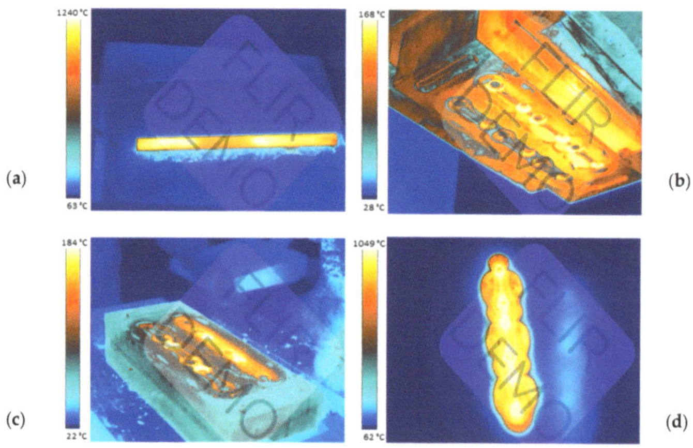

Figure 3. View with the temperature distribution during forging: (**a**) the charge lying on the feeder; (**b**) the temperature results for upper tools; (**c**) the results of lower tools; and (**d**) the temperature of forging.

The temperature measured by the pyrometer of the heater was 1320–1340 °C. The temperature of the charge leaving the heater is about 10–20 °C higher than that measured when the charge is lying on the feeder and waiting to be received by the operator. In turn, the working temperature of the tools is 200–250 °C. The heated charge material is fed into the roughing pass, in which roughing forging takes place by way of two blows, and next, in the finishing pass, one blow is performed. The final shape of the product is obtained as a result of hot trimming of the forging. A certain problem is the cooling off of the tools during forging, which is caused by the small volume of the deformed material with respect to the mass of the dies and the insufficient energy provided to the tools as a result of deformation and its work being exchanged into heat. In one forging cycle, the tools are in contact for a maximum of 0.5 s, whereas during the remaining time, the tools cool down as a result of radiation, conduction, and, to a small extent, convection with the environment. This is an issue which, in the case of process automatization, should also be solved in order to ensure prolonged work of the robots, without the necessity of frequent heating of the tools. Preliminary plans include the use of fast induction heating [49].

To sum up the performed analyses referring to the current technology, we should state that the production of a forging directly from a bar causes the formation of excessive flash, which is disadvantageous both with respect to die durability and a large material loss. Additionally, forging in a manual system causes a lack of stability and repeatability in the process, a decreased efficiency, and also translates to a lowered quality of the forged items and the formation of forging defects. For this reason, in order to improve the current technology, changes and solutions were introduced consisting of the development and design of a new tool construction with the consideration of forging in a robotized system. One of the implemented solutions is the use of an additional rolling pass for a preliminary re-forming of the material and the use of a longer charge in order to ensure the so-called tickworm, enabling the grip and holding of the charge/forging by the manipulators' grippers during the robotized forging process. To the current length of the input material in the form of a cylindrical bar, taking into account three times the length of a single forging together with the technological allowance, an allowance should be applied on both sides (about 20–30 mm each) for the so-called forceps, which will allow robots to hold the rod on both sides while forging. Additionally, a decision was made to reduce the charge diameter from 20 mm to 18 mm, which should translate to smaller material losses and lowered energy–force parameters. Introducing automatization and robotization of the currently realized technology should make it possible to solve the above problems and bring measurable benefits as well as improve the current forging process.

3.2. An Ideal Slug Forging and Rolling Pass

Before designing the rolling pass, we should first determine the ideal slug forging. To that end, we should calculate the section areas of the forging together with the flash in the particular areas. As the whole consists of three identical forgings, it is enough to calculate the section areas of one forging and next reflect it on the others. The assumed initial data are the die model, from which we cut out the body of the forging together with the flash, 10 mm wide, which is a sufficient value in the aspect of an easy-to-fill (during forging) shape of the connecting rod as well as a small height in reference to the width. On this basis, the particular section areas of the isolated body of the fogging were determined with respect to the length (Figure 4).

The diagram of the sections presented this way was smoothed so that it would be possible to use the correlated ideal slug forging for the preparation of the rolling pass (Figure 4c). With the data referring to the ideal slug forging, a rolling pass for a round bar, 14 mm in diameter, was designed. It was assumed that it would be a closed rolling pass, that is, of an elliptical shape. In such a case, according to the literature, the width of the pass b (the larger diameter of the ellipsis) should correspond to 3/2 of the height h (the smaller diameter of the ellipsis). The complete 3D model used to prepare the roller is shown in Figure 4d. Next, as a result of further research, a roller with a magazine was designed, which served to reduce the amount of energy needed for the deformation in the rolling pass. However, after the simulation, this solution was abandoned because there were no noticeable differences between the rollers in both versions, whereas the use of a magazine would involve unnecessary milling. After being connected to the lower and upper die (Figure 5), the roller obtained its final shape.

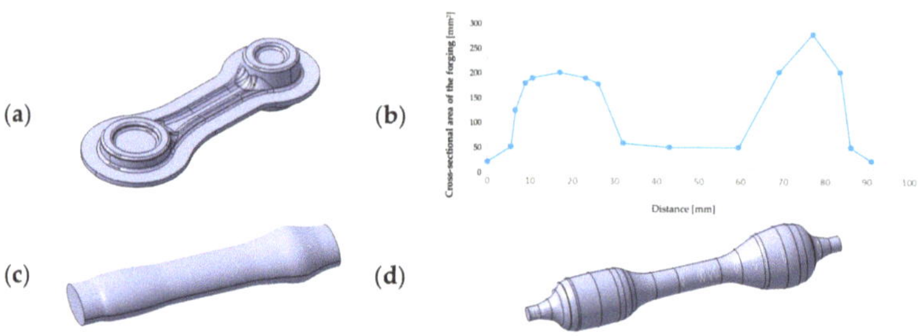

Figure 4. View of (**a**) a forging model with the predicted flash, (**b**) a diagram of the forging's sections, (**c**) a correlated ideal slug forging, and (**d**) a 3D model used to prepare the roller.

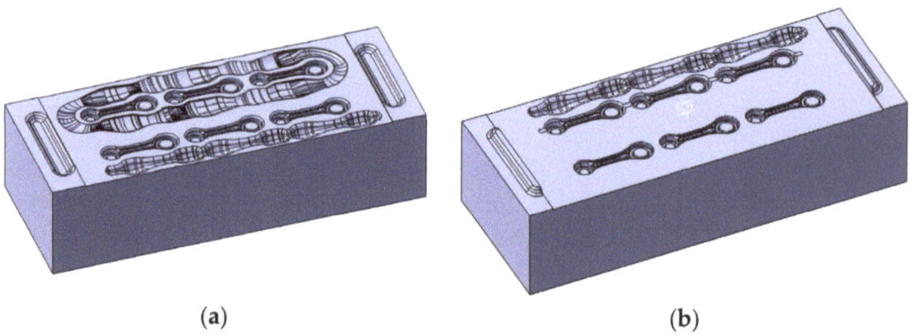

Figure 5. A 3D model of (**a**) the upper die with the roller and (**b**) the lower die with the roller.

In order to avoid the flash moving between the dies, in the area of its possible occurrence, the gap between the dies, 1.6 mm high, was expanded up to the border of the locks.

3.3. *Modeling of the Forging Process with Rolling*

The forging process of connecting rod-type forging on a robotized station will be realized with the application of an additional rolling operation, which was developed in order to reduce the diameter of the charge material from 20 mm to 18 mm. Special attention in the simulation was paid to the rolling operation, for which a dozen or so variants were created. During the deformation process, the tips of the bar were not deformed as they were to be held by the robots. The conditions assumed in the calculations of the forging process on the automatized station were exactly the same as those for manual forging, only the length of the bar was larger, and the shape of the die was adjusted in a way that would enable a non-collision placement of the robot's grippers between the upper and lower die. Also, in the roughing operation, one blow of the hammer was applied. The conditions in the calculations were assumed for a process of forging in three-pass dies. The number of blows: RX—the rolling pass—one blow; 1X (the roughing pass) with the energy of 10.5 kJ—one blow; 2X (the finishing pass): 6.2 kJ—one blow. The forging temperature and the charge temperature was 1320 °C. The cycle time was 12 s divided into cooling 6 s + forging in three operations. The machine was a hydraulic hammer with 16 kJ of energy. The tool temperature was 250 °C. The lubrication was water with graphite. The heat exchange was an average of 10 kW/(m^2·K). The simulation results for a Ø 18 mm diameter bar are presented in Figure 6 (models A and B).

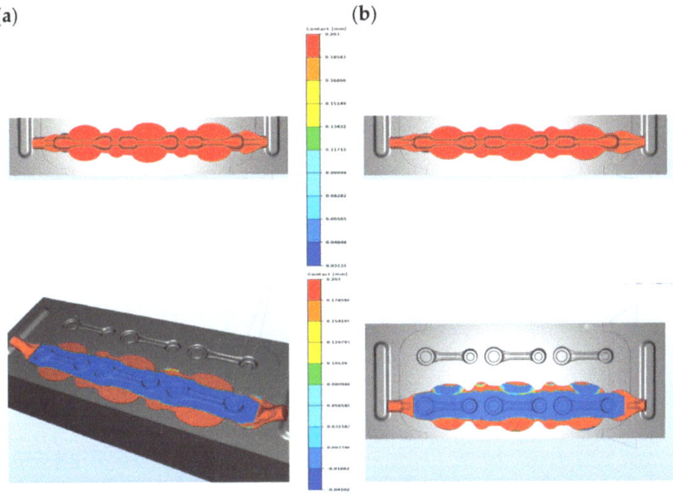

Figure 6. Contact with the tools in the pre-roughing operation, a 14 mm diameter bar: (**a**) model A—1.2 mm opening and (**b**) model B—1 mm opening.

For models A and B, we can see certain differences in the construction of the tools in the pre-roughing operation. Modification A has no bridge for the flash, whereas, in modification B, we can see the shape of a bridge (the color red denotes lack of contact with the die). Modification A has stronger friction forces on the flash, which can bring a more advantageous result in successive forging operations. The shape of the pass was previously filled, so a 1.2 mm opening was used, whereas, in modification B, a 1 mm opening was chosen (Figure 6b). The difference in the dimension of the forging measured in the direction of the hammer's stroke is in favor of solution A, as it is 0.2 mm bigger (Figure 6a). The results of the simulation in the roughing operation show that, regardless of the shape of the used flash, the filling of the pass is fully realized for a 14 mm diameter bar. For this reason, further works were initiated in order to perfect the technology, with the final assumed solution from model A. The approximate forming forces and the forging energy are shown in Figure 7 for the pre-roughing operation. The energy was about 5 kJ and the forging force was maximally 250 tons.

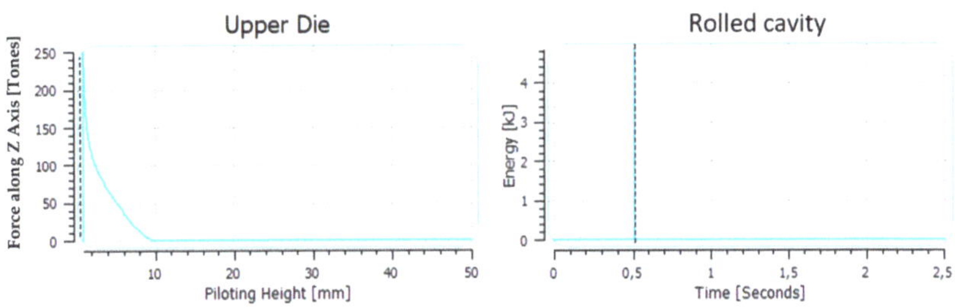

Figure 7. Force and energy of the blow—pre-roughing operation, variant B, a 1 mm opening.

A comparison was also made of the forces and energies of formation in the manual and automated forging technology in the roughing and finishing operation (Figure 8). In the roughing operation in the manual forging process, the hammer performs two blows, whereas in the case of adding the rolling operation and the application of a smaller diameter bar, i.e., 18 mm, we need one blow with the energy of about 10 kJ.

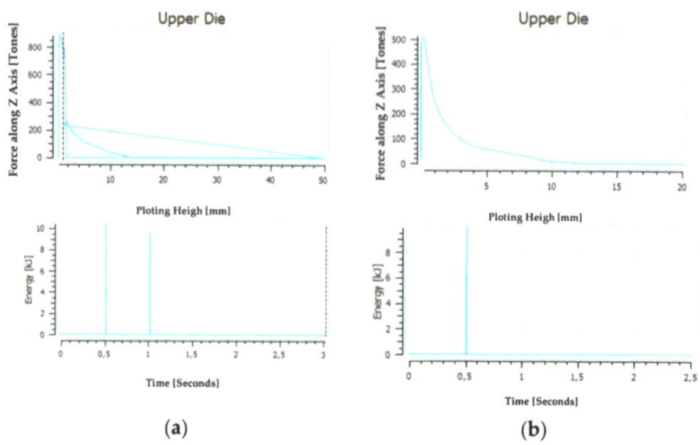

Figure 8. Comparison of the force–energy parameters for forging technology: (**a**) manually and (**b**) robotization.

The forming force can be lower in the case of the technology in three operations, and, for exemplary results, it equals about 500 tons, previously being over 800 tons. In turn, the energy needed to deform the detail in the finishing operation is lower in the simulation and equals 4.5 kJ, while in the case of manual forging, it is over 6 kJ. However, the force increased from 700 to 800 tons.

During the process, the temperature increases as a result of a change in the plastic deformation work into heat and, in the roughing operation, it equals about 945–1145 °C, which is presented in Figure 9. The plastic deformation in this operation equals maximally about 5—the highest takes place in the area of the changes in the cross-section.

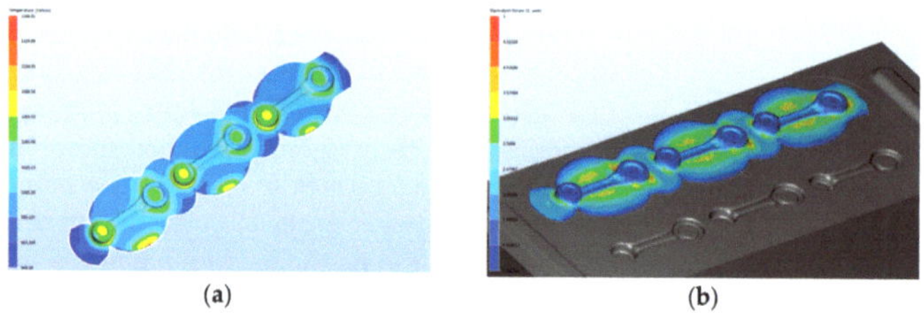

Figure 9. Results of the simulation after the roughing operation—distributions of (**a**) the temperature field and (**b**) the deformations.

The temperature at the end of the forging process after the finishing forging equals about 890–1100 °C (Figure 10). The highest temperature is on the side of the upper die. On the cross-section, we can see a big difference in the temperature field distribution.

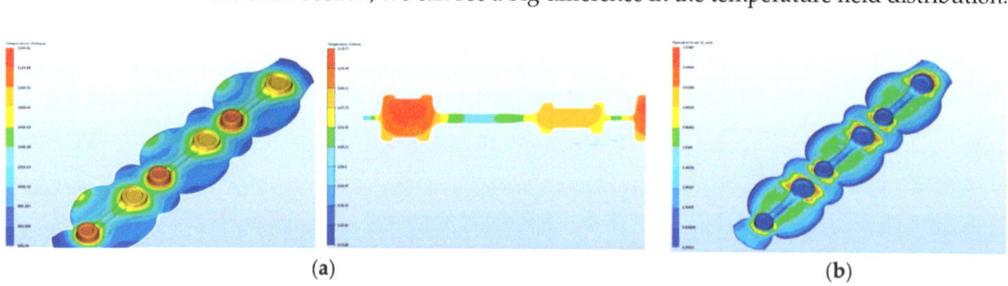

Figure 10. Results of the simulation after the finishing operation—distributions of (**a**) the temperature field and (**b**) the deformations.

The maximal plastic deformations increase only slightly. The deformation process runs properly. The formed lap occurs on the flash beyond the detail and does not affect the quality of the obtained products.

A detailed analysis was also made of the changes in the grain size during the forging in the newly elaborated robotized technology, in order to provide a possibility to model the microstructural changes during deformation and thus verify the usefulness of numerical modeling for such tasks connected with a microstructural analysis. The initial grain size was assumed at the level of 45 μm, based on the performed microscopic tests and analyses. In the QForm program, the mean grain size (its diameter in μm) is defined as follows:

$$d_{\mu m} = 1000\sqrt{10^{-\frac{ASTM+2.95}{3.32}}} \tag{1}$$

where ASTM can be read from the table below [38] (Table 1).

Table 1. Grain size referred to ASTM.

ASTM	15	14	13	12	11	10	9	8	7	6	5	4	3	2	1	0
Grain size (μm)	2	3	4	6	8	11	16	22	32	45	63	90	127	180	254	360

In turn, Figure 11 shows the grain size distributions in the final phase of the rolling operation, with a full pocket right after the forging process and 2 s after the forging (this is the time when the forging is replaced into the successive pass of the roughing forging operation).

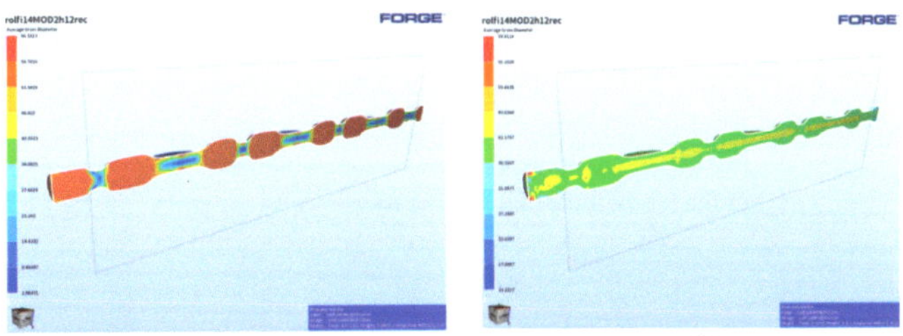

(**a**) (**b**)

Figure 11. Results of the simulation for the rolling operation: (**a**) the final deformation phase and (**b**) 2 s after the rolling process.

Based on the presented grain size distributions in the final phase of the process, we can observe that the largest mean grain diameters equaling 38–41 μm are localized in the areas of the forging with the highest volume, that is where the material was the least deformed. In areas of larger deformations, the grain is much smaller (Figure 8b). The case is similar for the roughing forging operation, where the deformation is at the level of 2–4, which means the grain size, as a result of microstructure reconstruction (dynamic recrystallization), is at a similar level (Figure 12).

Figure 13 shows the results of the forging simulation for the fishing operation, which included the distributions of deformation, temperature changes, and grain size right after the forging and 3 min after the end of the process.

We can notice that, as a result of the dynamic processes, the recrystallization did not take place in the analyzed time frame in the whole volume of the forging (Figure 12c), with respect to the deformation distribution for the forging in the final forging phase. In turn, we can see that, in the case of a longer time (Figure 13d,e), it has no significant effect on the grain growth. Moreover, the obtained modeling results for the time of 180 s after forging can be compared to the results of the microstructural tests for the forging after forging on the robotized station, because, as we can see, throughout a longer period of time, no significant changes in the microstructure can be observed. The mean grain size obtained from the numerical modeling equals about 35–40 μm (6 according to ASTM).

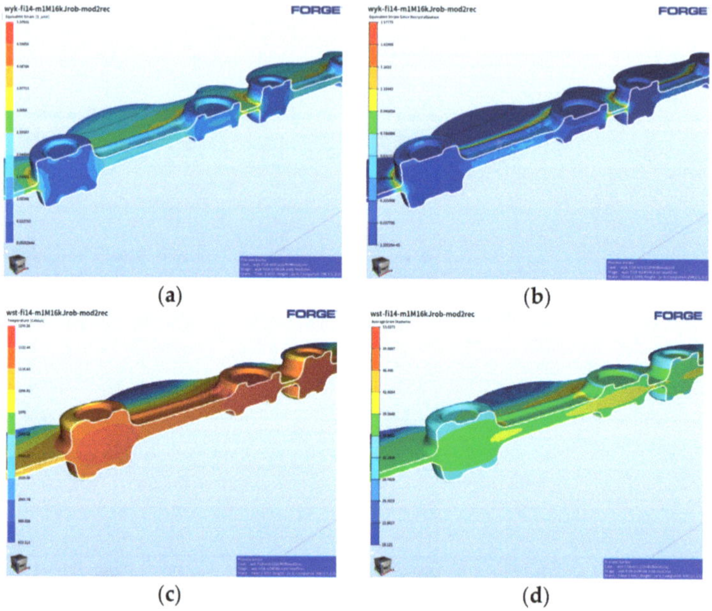

Figure 12. Results of the simulation for the roughing operation: (**a**) the total deformation distribution, (**b**) the deformation distribution only in the roughing forging (since recrystallization), (**c**) distributions of the temperature filed in the forging, and (**d**) grain size distributions.

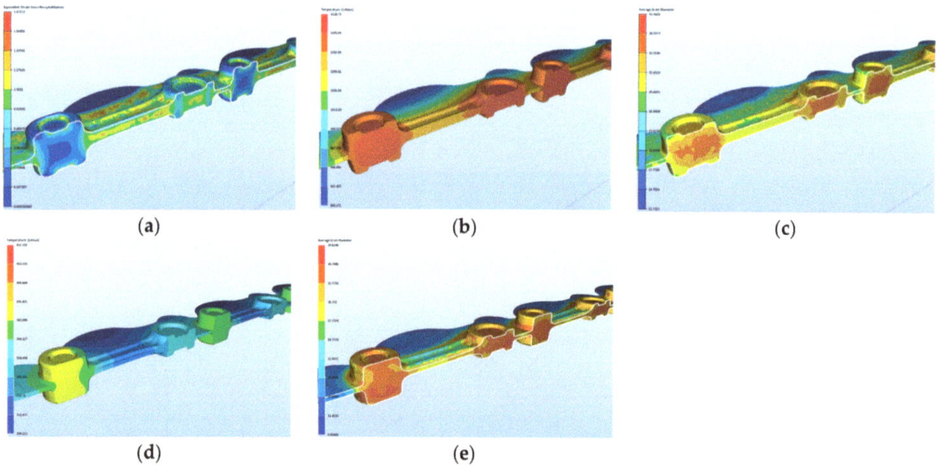

Figure 13. Results of the simulation for the finishing operation—the final phase of deformation: (**a**) deformation distributions, (**b**) temperature field distributions, (**c**) grain size distributions for this case, (**d**) temperature changes 3 min after the forging, and (**e**) grain size distributions for an analogical process.

The presented results with the grain size, plastic deformation, and temperature distributions demonstrate that, in reference to the results for the roughing operation, the changes in the grain size and plastic deformations are not that big, which results from the fact that the finishing operation is in a sense an operation of calibration. And so, the forging does not undergo large deformations, which makes the effect of the time elongation on the gain size and the plastic deformations small.

4. Trials under Industrial Conditions and Numerical Modeling Verification

During the robotization of the forging process, it was decided that, from the moment of placing the heated material onto the die to the moment of it being moved to the hot trimming operation, the formed element would be held on both sides by the robots' grippers, which would be moving synchronically during all the forging operations (Figure 14a). This would ensure stable support in two points, which would prevent the material from sliding out. The three robots assigned for the work would be performing the following tasks.

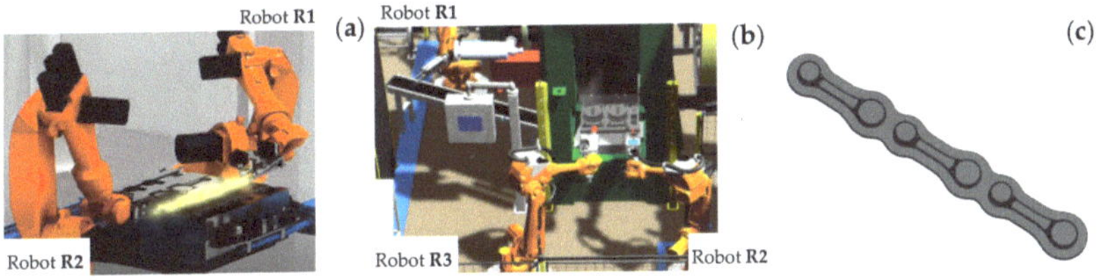

Figure 14. View of (**a**) an idea of the simultaneous work of both robots, (**b**) the simulation on the robotized station in the RobotStudio program, and (**c**) the CAD model of forged elements, the so-called "leaves", with single forgings.

Robot R1 supports the heated bar, which has been moved out of the induction heater, and places it above the lower die in such a way that the other end can be gripped by robot R2. Both robots manipulate the forging during the forging process, after which robot R1 releases its grip and the further manipulation takes place only by robot R2, which intercepts the forgings connected by the flash and next, places them in the trimming press tool through the side window. Figure 14b presents the preliminary forging tests performed on the robotized station.

In order to verify the properness of the forgings produced on the robotized seat, measurements of selected forged elements were made. In the first place, 3D scanning was carried out of the test batch of the connecting rod forgings collected from the forging process before the trimming. Presented below are representative scanning results. A cloud of points in the form of a triangle grid was obtained, which, after the measurement data equalization, was analyzed by means of the GOM software (version 2019). Figure 15 shows exemplary 3D scanning results in the form of a colored map of deviations for one leaf randomly selected from the whole series.

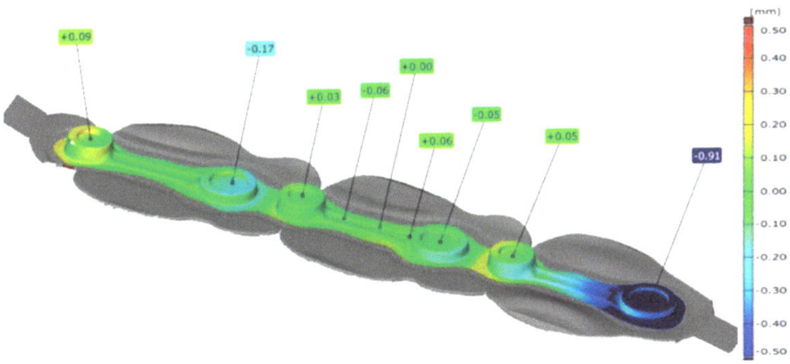

Figure 15. Results of 3D scanning of the K22 forging.

Analyzing the data shown in Figure 15, we can notice deformation of the forging's leaf, which is formed as a result of the forging process. We can see that, in the central part, which was used to equalize the data, the shape deviation is the smallest. In the case of the seat on the left, a slight deformation of the forging occurs. In the case of the forging forged on the right seat, we can see deformation of the external part of the forging at the level of −0.91 mm. Such a bend should not significantly affect the dimension–shape precision of single forgings. Nonetheless, detailed measurements of six selected leaves were made: two from the beginning of the process, two from the middle, and two from the end of the technological trials, and the obtained results are presented in Table 2.

Table 2. Results of the measurements of selected geometrical features for 6 randomly chosen forged elements (for example, 1B denotes 1 forged leaf, 2—the number of a single connecting rod forging).

	The First Attempt				
	Dd (mm) 30 ± 0.03	**Dm (mm)** 15 ± 0.2	**Tm (mm)** 12.5 ± 0.15	**Td (mm)** 17.3 ± 0.15	**Deviation (mm)** Max 0.3
1A	30.02	15.08	12.71	17.22	0.099
1B	29.99	15.07	12.72	17.24	0.048
1C	29.98	15.08	12.74	17.14	0.144
2A	29.99	15.04	12.68	17.21	0.064
2B	29.98	15.06	12.70	17.25	0.080
2C	30.03	15.07	12.72	17.20	0.124
	The second attempt				
1A′	30.01	15.07	12.58	17.22	0.079
1B′	30.02	15.04	12.63	17.26	0.089
1C′	29.98	15.04	12.58	17.23	0.162
2A′	29.99	15.05	12.58	17.23	0.029
2B′	29.98	15.06	12.67	17.28	0.060
2C′	30.02	14.84	12.61	10.25	0.127
	The last attempt				
3A″	29.99	15.05	12.60	17.19	0.028
3B″	29.98	15.06	12.61	17.25	0.075
3C″	30.03	15.04	12.59	17.20	0.180
3A″	30.01	15.07	12.58	17.20	0.029
3B″	30.02	15.06	15.62	17.27	0.082
3C″	29.98	15.06	12.60	17.23	0.039

As we can notice, the obtained measurement for the selected geometrical characteristics (Dm—the diameter of the small mesh; Dd—the diameter of the big mesh; Tm—the thickness of the small mesh; Tg—the thickness of the big mesh; and D—the joggle), are basically within the assumed dimension tolerances, which shows that the forging process in the developed robotized system is appropriate with respect to the dimension–shape accuracy of the forgings. Only for the first two leaves, the thickness results of the small mesh demonstrated that it was slightly higher, yet for the consecutive randomly selected elements, we can see that the process has stabilized, and all the geometrical features are in the tolerance field.

Additionally, in order to verify the performed preliminary forging tests in the robotized system, the initial material, i.e., steel 13CrMo4-5 (1.7335), was examined. The steel is chrome–molybdenum steel, containing 0.70–1.15% Cr and 0.4–0.6% Mo, making it one of the most noble and durable materials. Chromium steel has a relatively low carbon content compared to other related grades that are also used for high-temperature applications. The performed microscopy tests were carried out on a light microscope for microstructure verification. Figure 16 shows the results for the microstructure of the material as-delivered steel 1.7335.

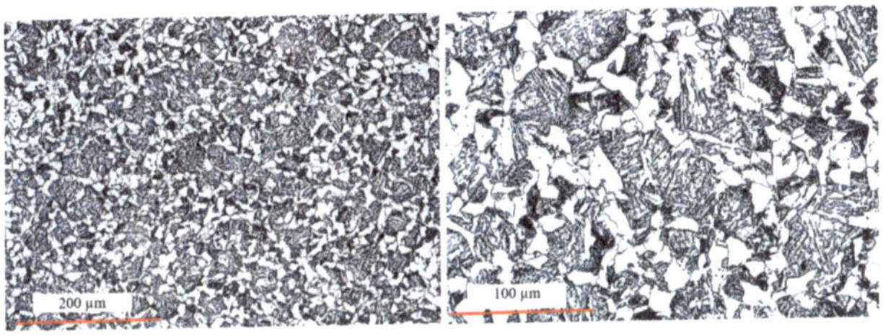

Figure 16. Microstructure of steel 1.7335—initial material for a connecting rod forging.

The microstructure of steel 16MnCr5 is a typical low-carbon ferritic–pearlitic microstructure with visible fine Fe_3C precipitates in the form of coagulated particles, not bound in the form of pearlite. It has a fine-grained structure. The measured hardness for this material as delivered equaled about 218 HV. In turn, for a randomly selected forging (from the middle part of the technological trials), metallographic tests as well as microstructural and grain size analyses were performed (in order to confront the results with those of numerical modeling). From the forging, three samples for each test were cut out. Figure 17 shows the areas of the shearing line for the samples for micro-section preparation, on the surface of which microstructural examinations were made as well as grain size evaluation and hardness measurements HV1 were performed.

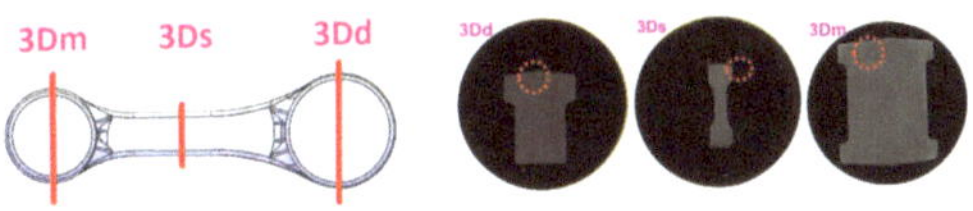

(**a**) (**b**)

Figure 17. Photographs of (**a**) the areas where the examined connecting rod was cut for the metallographic test samples for the connecting rod and (**b**) the appearance of the microsections from the cut-out samples and the mounted ones.

Figure 18 presents the microstructure test results for selected areas from the forging together with the measurement of the grain size. In turn, Table 3 shows the collective results of the microstructural analysis for a connecting rod forging. The microstructure was revealed by way of etching with a 2.5% Nital reagent and the microstructure observations were performed on the laser microscope Keyence VHX 6000 with magnifications of 100× and 200×. The determination of the grain sizes in the microstructures of the analyzed connecting rod was made by the secant method, with the use of the specialized software of the microscope.

Figure 18. *Cont.*

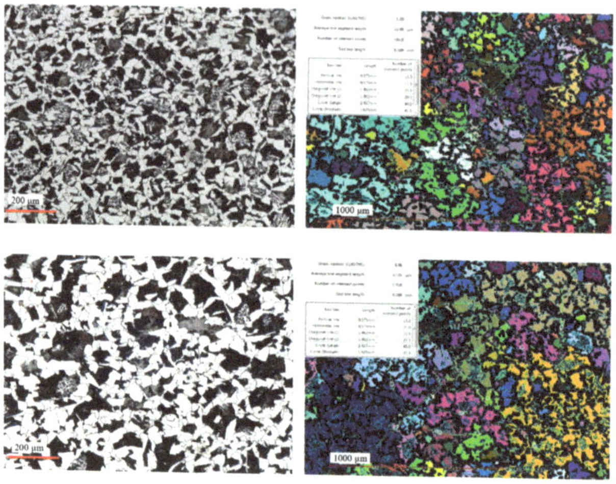

Figure 18. Results of the microstructural tests for selected areas from a randomly collected forging of a connecting rod: samples 3-Dd, 3-Ds, 3-Dm.

Table 3. Results of microscopic tests and hardness measurements.

Sample	Microstructure	Grain of Former Austenite PN-EN ISO 643:2020 [49]	Hardness HB	Decarburization Zone
3-Dd	ferrite + pearlite, ferrite present on grain boundaries	4.5	175	None
3-Ds	ferrite on grain boundaries + pearlite + bainite	5	175	None
3-Dm	bainite + ferrite on grain boundaries + traces of pearlite	5	190	None

The obtained test results referring to the microstructure confirm that introducing process robotization into the current technology has no negative effect on the microstructural changes, as the grain size, as well as the hardness and microstructure, are according to the requirements. In turn, referring to the grain size results obtained from the technological trials to FEM, we can state that the grain size is one grade lower (5 according to ASTM, that is about 65 µm) with respect to the modeling results (6 according to ASTM, that is 40 µm). On this basis, we can state that automatizing the current forging process realized in a multiple system through a transfer from a manual to an automatized process brings only advantageous aspects (dimension–shape precision and microstructure) as well as additionally ensures stability and repeatability of the production process. This said, before introducing the newly developed forging technology into series production, further tests and investigations should be carried out under industrial conditions.

5. Conclusions

The study presents the results referring to an improvement as well as the possibilities of introducing robotization into the currency-realized technology of precision forging on a hammer to produce connecting rod forgings with the use of mainly numerical modeling. Based on the analysis of the state of the art, it was proven that the available literature provides hardly any thorough studies or solutions for robotized forging seats; what is more, robotization cannot be treated as a universal solution. For this reason, the article has proposed a solution of optimizing the presently realized process of forging connecting rod forgings with the consideration of the aspects of robotization, e.g., by way of developing an additional rolling pass, reducing the diameter and mass of the charge material, and holding the charge/slug forging by robots' grippers. The investigations included a complex analysis of the present forging technology to determine the key areas requiring improvement/changes and next designing a new tool construction and modeling the technology by means of the calculation package Forge 3.0 NxT through multi-variant numerical simulations of the newly developed forging process. The obtained results have been verified under industrial conditions and correlated with respect to the forgings obtained in the technology realized so far. The attempts at introducing robotization aimed mostly at making the present manual process stable and repeatable, which would increase the quality of the forgings and at the same time reduce the rejection rate as a result of eliminating the errors produced by the operator. The obtained results of technological tests have confirmed that the introduced and then implemented changes were proper and brought measurable benefits. Both the dimension–shape accuracy and the hardness and microstructure are according to the requirements. The additional rolling pass reduced the forging forces and energy by over 30% and decreased the number of hammer blows by one, with respect to the manual process, which should also translate to an increased hardness of the forging instrumentation. This said, in order to ultimately verify the elaborated solution, it is necessary to perform further examinations and tests including a much longer operation time. It should also be emphasized that the presented results point to a big potential of the use of numerical modeling methods, as it currently enables full analysis of the process (including simulation of the microstructure development) as well as improvement of the given forging process.

Author Contributions: Conceptualization, Ł.D., M.H. and S.P.; methodology, M.H., A.B. and T.S.; software, S.P.; validation, Ł.D., A.B. and A.M.; formal analysis, M.H. and Ł.D.; investigation, M.H., Ł.D., A.M. and A.B.; resources, A.M.; data curation, Ł.D. and T.S.; writing—original draft preparation, M.H., Ł.D. and T.S.; writing—review and editing, M.H., Ł.D. and S.P.; visualization, A.M. and A.B.; supervision, M.H.; project administration, Ł.D. and T.S. All authors have read and agreed to the published version of the manuscript.

Funding: The research was carried out as a result of the project no. POIR.01.01.01-00-0537/21.

Institutional Review Board Statement: Not applicable.

Informed Consent Statement: Not applicable.

Data Availability Statement: Data are contained within the article.

Conflicts of Interest: Authors Łukasz Dudkiewicz, Adrian Miżejewski and Tatiana Szymańska are employed by Schraner Polska. The remaining authors declare that the research was conducted in the absence of any commercial or financial relationships that could be construed as a potential conflict of interest.

References

1. Magesh, K.; Ankush, K.B. A Review Paper on Design Analysis of Connecting Rod. *Int. J. Recent Innov. Trends Comput. Commun.* **2020**, *5*, 479–481.
2. Godara, S.S.; Nagar, S.N. Analysis of frontal bumper beam of automobile vehicle by using carbon. *Mater. Today Proc.* **2020**, *26*, 2601–2607. [CrossRef]
3. Sathish, T.; Kumar, S. Modelling and analysis of different connecting rod material through finite element route. *Mater. Today Proc.* **2020**, *21*, 971–975. [CrossRef]
4. Gronostajski, Z.; Hawryluk, M. The main aspects of precision forging. *Arch. Civ. Mech. Eng.* **2008**, *8*, 39–55. [CrossRef]
5. Altan, T. *Cold and Hot Forging Fundamentals and Application*; Ohio ASM International: Ohio, MI, USA, 2005.
6. Banaszek, G.; Stefanik, A. Theoretical and laboratory modelling of the closure of metallurgical defects during forming of a forging. *J. Mater. Process. Technol.* **2006**, *177*, 238–242. [CrossRef]
7. Hawryluk, M.; Rychlik, M. An implementation of robotization for the chosen hot die forging process. *Arch. Civ. Mech. Eng.* **2022**, *22*, 119. [CrossRef]
8. Belchiora, J.; Guillo, M.; Courteille, E.; Maurine, P.; Leotoing, L.; Guines, D. Off-line compensation of the tool path deviations on robotic machining: Application to incremental sheet forming. *Robot Comput. Integrat. Manuf.* **2013**, *29*, 58–69. [CrossRef]

9. Mitchell, G.; Kramer, A. *Forging Operations—Machine Forging, Forging Dies and Special Forging Operations*; Read Books Ltd.: Redditch, UK, 2013.
10. Van Doren, M.J.; Slocum, A. Design and implementation of a precision material handling robot control system. *Int. J. Mach. Tools Manuf.* **1995**, *35*, 1003–1014. [CrossRef]
11. SMS Group GmbH. Available online: https://www.sms-group.com (accessed on 17 October 2023).
12. SCHUNK SE & Co. KG. Available online: https://schunk.com/pl_pl/systemy-chwytakowe/ (accessed on 17 October 2023).
13. Ferrari, P.; Flammini, A.; Vitturi, S. Performance analysis of PROFINET networks. *Comput. Stand. Interfaces* **2006**, *28*, 369–385 [CrossRef]
14. Honczarenko, J. *Industrial Robots Construction and Application*, 2nd ed.; Scientific and Technical Publishing Houses: Warsaw, Poland, 2010.
15. Yin, G.; Zhu, Z.; Gong, H.; Lu, Z.; Yong, H.; Liu, L.; He, W. Robotics and Computer–Integrated Manufacturing Flexible punching system using industrial robots for automotive panels. *Robot. Comput. Integr. Manuf.* **2018**, *52*, 92–99. [CrossRef]
16. Fu, X.; Bao, Q.; Xie, H.; Fu, X. Diffusion of industrial robotics and inclusive growth: Labour market evidence from cross country data. *J. Bus. Res.* **2021**, *122*, 670–684. [CrossRef]
17. *ISO 10360-4:2000*; Geometrical Product Specifications (GPS)–Acceptance and Reverification Tests for Coordinate Measuring Machines (CMM)–Part 4: CMMs Used in Scanning Measuring Mode. ISO: Geneva, Switzerland, 2000. Available online: https://hdl.handle.net/11696/74641 (accessed on 2 March 2023).
18. Dieter, G.E.; Kuhn, H.A.; Semiatin, S.L. *Handbook of Workability and Process Design. Materials Park*; ASM International: Geneva, Switzerland, 2003.
19. Ponthot, J.P.; Kleinermann, J.P. Optimisation methods for initial/tool shape optimisation in metal forming processes. *Int. J. Veh. Des.* **2005**, *39*, 14–24. [CrossRef]
20. Bonte, M.H.A.; Van den Boogaard, A.H.; Huétink, J. An optimisation strategy for industrial metal forming processes. *Struct. Multidiscip. Optim.* **2008**, *35*, 571–586. [CrossRef]
21. Lange, K. Modern metal forming technology for industrial production. *J. Mater. Process. Technol.* **1997**, *71*, 2–13. [CrossRef]
22. Jolgaf, M.; Hamouda, A.M.S.; Sulaiman, S.; Ahmad, M.M.H.M. Development of a CAD/CAM system for the closed-die forging process. *J. Mater. Process. Technol.* **2003**, *138*, 436–442. [CrossRef]
23. Srinivasan, N.; Ramakrishnan, N.; Venugopal Rao, A.; Swamy, N. CAE for forging of titanium alloy aero-engine disc and integration with CAD–CAM for fabrication of the dies. *J. Mater. Process. Technol.* **2002**, *124*, 353–359. [CrossRef]
24. Li, S.Y.; Cheng, S.Y. Design Optimization for Cold Forging by an Integrated Methodology of CAD/FEM/ANN. *Adv. Mater. Res.* **2010**, *97–101*, 3281–3284. [CrossRef]
25. Neugebauer, R.; Bräunlich, H.; Scheffler, S. Process monitoring and closed loop-controlled process. *Arch. Civ. Mech. Eng.* **2009**, *9*, 105–126. [CrossRef]
26. Ou, H.; Wang, P.; Lu, B.; Long, H. Finite element modelling and optimisation of net-shape metal forming processes with uncertainties. *Comput. Struct.* **2012**, *90–91*, 13–27. [CrossRef]
27. Wilby, A.J.; Neale, D.P. Defects Introduced into Metals during Fabrication and Service. *Mater. Sci. Eng.* **2009**, *3*, 48–75.
28. Wojtaszek, M.; Lisiecki, Ł.; Łukaszek-Sołek, A.; Korpała, G.; Zyguła, K.; Śleboda, T.; Jabłońska, M.B.; Prahl, U. Application of processing maps and numerical modelling for identification of parameters and limitations of hot forging process. *Arch. Civ. Mech. Eng.* **2023**, *23*, 240. [CrossRef]
29. Park, J.J.; Rebelo, N.; Kobayashi, S. A new approach to preform design in metal forming with the finite element method. *Int. J. Mach. Tool Des. Res.* **1983**, *23*, 71–92. [CrossRef]
30. Hawryluk, M.; Ziemba, J. Application of the 3D reverse scanning method in the analysis of tool wear and forging defects. *Measurement* **2018**, *128*, 204–213. [CrossRef]
31. Bouissa, Y.; Bohlooli, N.; Shahriari, D.; Champliaud, H.; Morin, J.B.; Jahazi, M. FEM modeling and experimental validation of quench-induced distortions of large size steel forgings. *J. Manuf. Process.* **2020**, *58*, 592–605. [CrossRef]
32. Xu, Y.; Zhang, Y.; Zhuang, X.; Cao, Z.; Lu, Y.; Zhao, Z. Numerical modeling and anvil design of high-speed forging process for railway axles. *Int. J. Mater. Form.* **2021**, *14*, 813–832. [CrossRef]
33. Lu, B.; Ou, H. An efficient approach for trimming simulation of 3D forged components. *Int. J. Mech. Sci.* **2012**, *55*, 30–41. [CrossRef]
34. Kawka, M.; Kakita, T.; Makinouchi, A. Simulation of multi-step sheet metal forming processes by a static explicit FEM code. *J. Mater. Process. Technol.* **1998**, *80–81*, 54–59. [CrossRef]
35. Lee, S.; Quagliato, L.; Park, D.; Kwon, I.; Sun, J.; Kim, N. A New Approach to Preform Design in Metal Forging Processes Based on the Convolution Neural Network. *Appl. Sci.* **2021**, *11*, 7948. [CrossRef]
36. Krishna, R.H.; Jena, D.P. Analytical and numerical modelling of open-die forging process for elliptical cross-section of billet. *Measurement* **2019**, *134*, 855–865. [CrossRef]
37. Transvalor. Forge 2011 Documentation—Datafile Forge 3v75. Available online: http://www.transvalor.com/en/cmspages/forge-nxt.32.html (accessed on 17 October 2023).
38. Qform. *Qform 3D—Software for Simulation and Optimization of Metal Forming Processes and Metal Profile Extrusion*, Qform UK: Oxford, UK, 1991. Available online: http://www.qform3d.com(accessed on 17 October 2023).

39. Simufact Forming. Hexagon. Available online: https://www.simufact.com/simufactforming-forming-simulation.html (accessed on 17 October 2023).
40. SFTC. DEFORM-3D. Scientific Forming Technologies Corporation. Available online: https://www.deform.com/products/deform-3d/ (accessed on 17 October 2023).
41. Hawryluk, M.; Jakubik, J. Analysis of forging defects for selected industrial die forging processes. *Eng. Fail. Anal.* **2016**, *59*, 396–409. [CrossRef]
42. Hawryluk, M.; Dudkiewicz, Ł.; Marzec, J.; Rychlik, M.; Tkocz, R. Selected aspects of lubrication in die forging processes at elevated temperatures—A review. *Lubricants* **2023**, *1*, 206. [CrossRef]
43. Hawryluk, M.; Polak, S.; Rychlik, M.; Dudkiewicz, Ł.; Borowski, J.; Suliga, M. Possibilities of Measuring and Detecting Defects of Forged Parts in Die Hot-Forging Processes. *Materials* **2024**, *17*, 213. [CrossRef] [PubMed]
44. Sharma, S.; Sharma, M.; Gupta, V.; Singh, J. A Systematic Review of Factors Affecting the Process Parameters and Various Measurement Techniques in Forging Processes. *Steel Res. Int.* **2023**, *94*, 2200529. [CrossRef]
45. Šraml, M.; Stupan, J.; Potrč, I.; Kramberger, J. Computer-aided analysis of the forging process. *Int. J. Adv. Manuf. Technol.* **2004**, *23*, 161–168. [CrossRef]
46. Shirgaokar, M. Chapter 9: Methods of Analysis for Forging Operations. In *Cold and Hot Forging: Fundamentals and Applications*; ASM Technical Books: Geneva, Switzerland, 2005. [CrossRef]
47. Lange, K.; Cser, L.; Geiger, M.; Kals, J.A.G. Tool Life and Tool Quality in Bulk Metal Forming. *Proc. Inst. Mech. Eng. Part B J. Eng. Manuf.* **1993**, *207*, 223–239. [CrossRef]
48. Hawryluk, M.; Widomski, P.; Kaszuba, M.; Krawczyk, J. Development of new preheating methods for hot forging tools based on industrial case studies and numerical modeling. *Metall. Mater. Trans. A Phys. Metall. Mater. Sci.* **2020**, *51*, 4753–4764. [CrossRef]
49. *ISO 643:2020*; Steels—Micrographic Determination of the Apparent Grain Size. Polish Committee for Standardization: Warsaw, Poland, 2020.

MDPI AG
Grosspeteranlage 5
4052 Basel
Switzerland
Tel.: +41 61 683 77 34

Materials Editorial Office
E-mail: materials@mdpi.com
www.mdpi.com/journal/materials